AF361848

9th EASN International Conference on Innovation in Aviation & Space

9th EASN International Conference on Innovation in Aviation & Space

Editors

Spiros Pantelakis
Andreas Strohmayer

MDPI • Basel • Beijing • Wuhan • Barcelona • Belgrade • Manchester • Tokyo • Cluj • Tianjin

Editors
Spiros Pantelakis
University of Patras
Greece

Andreas Strohmayer
University of Stuttgart
Germany

Editorial Office
MDPI
St. Alban-Anlage 66
4052 Basel, Switzerland

This is a reprint of articles from the Special Issue published online in the open access journal *Aerospace* (ISSN 2226-4310) (available at: https://www.mdpi.com/journal/aerospace/special_issues/9th_EASN).

For citation purposes, cite each article independently as indicated on the article page online and as indicated below:

LastName, A.A.; LastName, B.B.; LastName, C.C. Article Title. *Journal Name* **Year**, *Volume Number*, Page Range.

ISBN 978-3-0365-4223-2 (Hbk)
ISBN 978-3-0365-4224-9 (PDF)

Contents

About the Editors

Spiros Pantelakis

Spiros Pantelakis is Director of the Laboratory of Technology and Strength of Materials of the Mechanical Engineering and Aeronautics Department at the University of Patras. He served as Chairman of the Department of Mechanical Engineering and Aeronautics from September 2009 to September 2013. In September 2007 to August 2010, he was an Executive Board member of the Research Committee of the University of Patras. Furthermore, between September 2010 and August 2013, he was Vice Chairman of the Executive Board of the Research Committee of the University of Patras. He is one of the founding members and, since 2008, has been Chairman of the European Aeronautics Science Network (EASN) Association. From 2006 to 2011, he was Representative of the European Aeronautics Academia in the plenary of ACARE, and from 2005 to 2011, he was Chairman of ACARE's Working Group on Human Resources and Chairman of the Board of Directors of the Greek Metallurgical Society from 2008 to 2011. Since 2015, he has been a member of the Board of Directors of the Hellenic Aerospace Industry. He has over 35 years' experience in the field of aerostructures and aeronautics materials, during which he has been involved in over 100 international aeronautics research projects—in many of them, as Scientific Coordinator. He is a member of the Editorial Board and Guest Editor-in-Chief for a number of international scientific journals and Reviewer for many international scientific journals. Furthermore, he has been author or served as Editor in numerous books published by various international publishing houses. He is the author of more than 250 scientific publications in international peer-reviewed journals and conference proceedings. He has been Chairman of some dozens of international conferences and workshops, and he is founder of the Conference Series ICEAF (International Conference of Engineering Against Failure).

Andreas Strohmayer

Andreas Strohmayer is professor of aircraft design at the Institute of Aircraft Design, University of Stuttgart, with a research focus on manned (hybrid) electric flight ("Icaré 2" and "e-Genius") and scaled UAS flight testing. He studied Aeronautical Engineering at TU Munich and graduated in 2001 under Dr.-Ing. in the field of conceptual aircraft design. From 2002 to 2008, he was director of Grob Aerospace in Mindelheim, Germany, responsible for the design, production and support of the Grob fleet of all-composite aircraft: the development of a four-seat aerobatic turboprop, a seven-seat turboprop, and the SPn business jet. From 2009 to 2013, he was program director for a 19-seater commuter aircraft project at Sky Aircraft in Metz, France, then VP Programs at SST Flugtechnik in Memmingen, Germany, setting up an EASA-approved design organization, holding the TC for a six-seater all-composite turbo-prop aircraft. He then left the industry to join University of Stuttgart in 2015, teaching aircraft design and promoting electric flight. In 2016, he became member of the Board of Directors of the European Aeronautics Science Network (EASN), and since 2019, he has been an EASN Chairman. In this position, he also represents academia in the ACARE General Assembly.

Editorial

Special Issue "9th EASN International Conference on Innovation in Aviation & Space"

Spiros Pantelakis [1,*] and Andreas Strohmayer [2,*]

[1] Department of Mechanical Engineering and Aeronautics, University of Patras, Panepistimioupolis Rion, 26500 Patras, Greece
[2] Department of Aircraft Design, Institute of Aircraft Design (IFB), University of Stuttgart, Pfaffenwaldring 31, 70569 Stuttgart, Germany
* Correspondence: pantelak@mech.upatras.gr (S.P.); strohmayer@ifb.uni-stuttgart.de (A.S.)

Citation: Pantelakis, S.; Strohmayer, A. Special Issue "9th EASN International Conference on Innovation in Aviation & Space". *Aerospace* **2021**, *8*, 110. https://doi.org/10.3390/aerospace8040110

Received: 12 April 2021
Accepted: 13 April 2021
Published: 14 April 2021

This Special Issue contains selected papers from works presented at the 9th EASN International Conference on Innovation in Aviation & Space, which was successfully held in Athens, Greece, between the 3rd and 6th of September 2019. The event included 9 keynote lectures and more than 360 technical presentations distributed in approximately 70 sessions. Furthermore, 40 HORIZON2020 projects disseminated their latest research results, as well as the future trends on the respective technological field. In total, more than 450 participants joined the 9th EASN International Conference.

In the present Special Issue, eight engaging articles are contained, with more than 1800 views each until now, related to aviation and space research. Slavik et al. [1] performed a multi-objective optimization in order to select a fixed propeller for a short take-off and landing (STOL) category aircraft. The aim was to achieve the highest possible performance with fixed propeller, i.e., high maximal horizontal and cruise speed, short take-off and high rate of climb; Pareto sets were implemented in order to select the optimal propeller. As a result, a high-performance power system with a low price (fixed pitch propeller) for STOL category aircraft could be designed. Plagianakos et al. [2] studied the effect of hot–wet storage aging on the mechanical response of a carbon fiber polyether ether ketone (PEEK)-matrix woven composite. A wide range of static loads and selected cyclic load tests on the interlaminar fatigue strength were performed, with the results providing a useful basis towards preliminary design with PEEK-based woven thermoplastic composites during service in aerospace applications. Capovilla et al. [3] designed a CFRP structural/battery array configuration in order to integrate the electrical power system with a spacecraft bus primary structure. The results indicated that, by implementing the designed configuration, more volume and mass was made available for the payload, compared with traditional, functionally separated structures employing aluminum alloys. The study of Khustochka et al. [4] proposed a novel method for a stable estimation of the engine performance parameters using a priori information about the engine, its mathematical model and expected performance, in view of fuzzy sets, and also about the measuring system and measuring procedure. A comparison of the proposed approach with traditional methods underlined its main advantage regarding its high stability of estimation in the parametric uncertainty conditions, while the proposed method can be implemented for matching thermodynamic models to experimental data, gas path analysis, as well as adapting dynamic models to the needs of the engine control system. In the work of Piscitelli et al. [5], a superhydrophobic coating for metallic substrates with a simplified and non-expensive method was developed, which could be employed as a usual paint able to prevent/reduce the formation of ice, especially on small aircraft. The surface properties and the wettability of the developed coating were investigated, with the authors concluding that the specific coating can be potentially employed as a passive anti-icing system for aeronautical applications. Kellerman et al. [6] investigated the potential of using existing aircraft surfaces as heat sinks for the waste heat of a (hybrid-) electric drive train. The results pointed out that surface heat exchangers can provide cooling power in the same order of magnitude as the waste heat expected from (hybrid-) electric drive trains for all sizes of considered aircraft. Dvirnyk et al. [7] assessed the serviceability limit of the blades of the axial compressor of a helicopter engine operating in a dusty environment, and showed that the gradual loss of the stall margin over time determines the serviceability limits of compressor blades. The serviceability limits defined by the authors could enable helicopter users to significantly reduce operating costs by extending the remaining useful life (RUL) of the engines operated in a desert environment. Finally, in the study of Biser et al. [8], the possibilities of coupled, analytical models for sizing electric propulsion systems were demonstrated by considering the example of a propulsion system developed in the frame of a relevant EU-funded project. The approach proposed by the authors was found to favor model accuracy and low computational effort. Potential technical solutions to decrease the influence of the DC power transmission system were also given.

The editors of this Special Issue would like to thank the authors for their high-quality contributions and for making this Special Issue manageable. Additionally, the editors would like to thank Ms. Linghua Ding and the Aerospace editorial team.

Conflicts of Interest: The authors declare no conflict of interest.

References

1. Slavik, S.; Klesa, J.; Brabec, J. Propeller Selection by Means of Pareto-Optimal Sets Applied to Flight Performance. *Aerospace* **2020**, *7*, 21. [CrossRef]
2. Plagianakos, T.S.; Muñoz, K.; Saenz-Castillo, D.; Mendias, M.M.; Jiménez, M.; Prentzias, V. Effect of Hot-Wet Storage Aging on Mechanical Response of a Woven Thermoplastic Composite. *Aerospace* **2020**, *7*, 18. [CrossRef]
3. Capovilla, G.; Cestino, E.; Reyneri, L.M.; Romeo, G. Modular Multifunctional Composite Structure for CubeSat Applications: Preliminary Design and Structural Analysis. *Aerospace* **2020**, *7*, 17. [CrossRef]
4. Khustochka, O.; Yepifanov, S.; Zelenskyi, R.; Przysowa, R. Estimation of Performance Parameters of Turbine Engine Components Using Experimental Data in Parametric Uncertainty Conditions. *Aerospace* **2020**, *7*, 6. [CrossRef]
5. Piscitelli, F.; Chiariello, A.; Dabkowski, D.; Corraro, G.; Marra, F.; Di Palma, L. Superhydrophobic Coatings as Anti-Icing Systems for Small Aircraft. *Aerospace* **2020**, *7*, 2. [CrossRef]
6. Kellermann, H.; Habermann, A.L.; Hornung, M. Assessment of Aircraft Surface Heat Exchanger Potential. *Aerospace* **2020**, *7*, 1. [CrossRef]
7. Dvirnyk, Y.; Pavlenko, D.; Przysowa, R. Determination of Serviceability Limits of a Turboshaft Engine by the Criterion of Blade Natural Frequency and Stall Margin. *Aerospace* **2019**, *6*, 132. [CrossRef]
8. Biser, S.; Wortmann, G.; Ruppert, S.; Filipenko, M.; Noe, M.; Boll, M. Predesign Considerations for the DC Link Voltage Level of the CENTRELINE Fuselage Fan Drive Unit. *Aerospace* **2019**, *6*, 126. [CrossRef]

Article

Propeller Selection by Means of Pareto-Optimal Sets Applied to Flight Performance

Svatomir Slavik, Jan Klesa * and Jiri Brabec

Department of Aerospace Engineering, Czech Technical University in Prague, Prague 121 35, Czech Republic;
svatomir.slavik@fs.cvut.cz (S.S.); j.brabec@fs.cvut.cz (J.B.)
* Correspondence: jan.klesa@fs.cvut.cz

Received: 30 November 2019; Accepted: 28 February 2020; Published: 5 March 2020

Abstract: Selection process of the propeller for short take-off and landing (STOL) category aircraft is described. The aim is to achieve the highest possible performance with fixed propeller, i.e., high maximal horizontal and cruise speed, short take-off and high rate of climb. These requirements are contradictory and so Pareto sets were used in order to find the optimal propeller. The method is applied to a family of geometrically similar propellers that are suitable for 73.5 kW (100 hp) piston engine designed for ultralight category aircraft with maximal take-off weight of 472.5 kg. The propellers have from two to eight blades, blade angle settings from 15° to 40° and diameter from 1.1 m to 2.65 m. Pareto frontier is designed for each pair of flight conditions, and the optimal propeller is selected according to these results. For comparison, the optimal propeller selection from the propeller family by means of a standard single-optimal process based on the speed power coefficient c_s is also used. Use of Pareto sets leads to considerable performance increase for the set of contradictory requirements. Therefore, high performance for a low price for the given aircraft can be achieved. The described method can be used for propeller optimization in similar cases.

Keywords: STOL aircraft; propeller; Pareto sets; propeller optimization

---

1. Introduction

Optimal propeller performance has been investigated from the beginning of aviation. The first scientific work on propeller aerodynamic optimization was performed by Betz and Prandtl [1]. A more precise method was described by Goldstein [2]. The theory of Betz and Prandtl was used by Larrabee for the aerodynamic design of the propellers with low loading [3]. This method is quite popular until today due to its simplicity and good results. It was later developed, e.g., by Adkins and Liebeck [4] or Hepperle [5]. All these methods can be used for the aerodynamic design of optimal propeller for given flight condition.

This paper is focused on the choice of optimal propeller from given family, i.e., group of propellers having identical blade shape and various diameter and number of blades for various flight conditions, i.e., requirements for the propeller are usually contradictory. Standard method for the solution of this problem by means of power speed coefficient c_s is described in [6–8]. This procedure is relatively simple and determines optimal diameter and blade angle for the group of propellers having the same blade shape and number of blades. This works only for a single operating point.

Therefore, if propeller performance should be maximal in several flight conditions, some multiple objective optimization method should be used. This brings difficulties in identifying the propeller with the best performance and thus the optimal solution. It is possible to use Pareto-optimal sets for this task. Its application is quite frequent. It can be used for the multi-objective optimization of ship propellers [9,10], optimization of UAS powerplant [11] or propeller optimization with manufacturing constraints [12]. Other possibilities are, e.g., gradient-based methods [13], genetic algorithms [14] or calculus of variations [15].

Multi-objective propeller optimization described in [9–15] is focused on the propeller design. This paper brings a new point of view. It combines the approach from [6–8] with multi-disciplinary optimization using Pareto-optimal sets. The optimal propeller is selected from the family of propellers, i.e., propellers having the same blade shape but different diameter and number of blades. The main advantage of this problem definition is the prevention of possible structural problems. Increase in the number of blades together with lower blade cord leads to higher efficiency, but it leads also to lower bending stiffness and strength (e.g., for the blade chord decreased to 50% of its baseline value, bending stiffness decreases to 12.5% and bending strength to 25% of its baseline values). To avoid this, optimal propeller is selected from the set similar to the approach in [7,8].

2. Problem Formulation

The task is to choose the optimal fixed-pitch propeller for a given aircraft and engine. The objectives are to maximize both cruise and maximal horizontal flight speed and at the same time minimize the take-off distance of the aircraft. These results are contradictory, so some sort of trade-off must be accepted. The method itself can be used for any propeller driven aircraft. Further, it is presented in the example of the light short take-off and landing (STOL) category aircraft. Development of this aircraft also was main motivation for the research in this branch. The aircraft was originally equipped with constant-speed propeller. The usage of the fixed-pitch propeller was motivated by the decrease in both system complexity and cost.

3. Methods

3.1. Pareto Sets of Flight Performance for Selection of Optimal Propellers

Multi-objective optimization by means of Pareto sets is based on search for boundary values of objective functions dependent on the same variable X of these functions. Values of objective functions are evaluated for every value of the independent variable X and plotted in the graph (see Figure 1). An element of Pareto set represents mutual relation of all considered objective functions for one given value of the independent variable X.

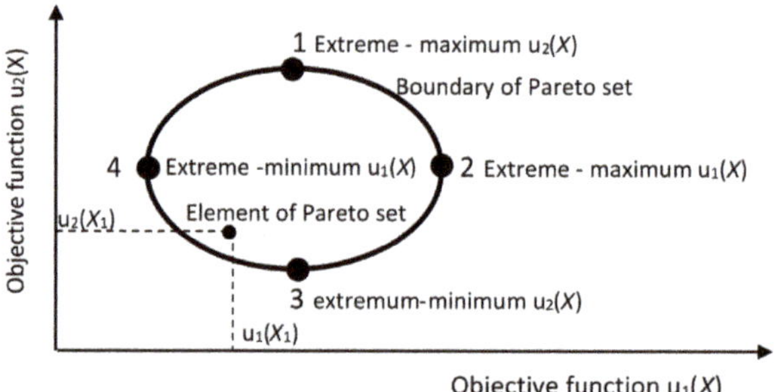

Figure 1. Scheme of Pareto set (redrawn according to [16]).

In the case of multi-objective optimization with two objective functions, Pareto sets have the form of plane graphs. One axis represents values of the first objective function and the second one values of the second objective function, as shown in Figure 1. Corresponding pairs of both functions (for the same variable values X) create Pareto set as the area of the graph. The set area limits are bound to a defined condition range of an independent variable for each function. The bound of Pareto sets makes outer limits of achievable values of the objective functions.

Peak points of the bound (Points 1, 2, 3, and 4 in Figure 1) depict extremes (minimum, maximum) of one and the other objective functions. Parts of the boundary between extremes match optimal values of variable X, because if the extremes of both these functions are simultaneously required (e.g., maximum 1 for $u_2(X)$ and maximum 2 for $u_1(X)$), the extreme is achievable only for the first or second function. The change on Pareto boundary when dropping under the extreme of the first function approximates in the optimal way the solution to the extreme of the second function (for each variable X, between 1 and 2, both functions reach their maximum values). All variables X on Pareto boundary between these two extremes (Pareto-optimal front) are optimal (the most appropriate) because it is not possible to decide which combination of both objective functions is better. None of these solutions is worse or better—the solutions are mutually non-dominant.

In the case that the extreme of one function is also the extreme of the second function (for one value of the variable X, both extremes are reached) then Pareto-optimal front passes at one point (e.g., extremes 1 and 2 are identified) and the multi-objective optimization is one optimal solution only.

The goal of optimization is to find a non-dominant solution that requires the functions to be in contradiction (conflict). A change of variable X for one function towards its required extreme value delays the second function from its required minimum/maximum. A more detailed description of multi-objective optimization by means of Pareto sets can be found in literature [16] or [17].

In the optimization selection of propellers from the family of propeller, the discrete points represent individual propellers. The propeller geometry comprises both continuously geometric parameters (mainly distribution of the chord length, twist and thickness of the airfoils used along the blade, the propeller diameter) and discrete parameter—the number of blades. The propeller aerodynamic characteristics represent the dependence of the thrust and power coefficients on the advance ratio $(c_T(\lambda), c_P(\lambda))$ that together with an engine power curve enables to set the available isolated thrust curve of the power unit.

Pareto sets are composed from final number of points (equal to the number of propellers in the family of propeller) of appropriate pairs of the contradictory flight conditions objective functions: [maximum horizontal flight speed–take-off distance], [maximum horizontal flight speed–maximum rate of climb], [take-off distance–maximum rate of climb]. The pair of [maximum horizontal flight speed (continuous engine regime)–maximum horizontal flight speed (cruise regime)] is also included.

The optimization selection of the propeller by means of Pareto-optimal sets consists of the following sequential actions:

1. Assessment of free flight aerodynamic characteristics of the airplane (lift curve $c_L(\alpha)$, lift-drag polar $c_L(c_D)$ without the propeller influence on the airplane drag in the relevant flight configuration of the flight performance.
2. Determination of propeller aerodynamic characteristics—thrust and power coefficients $c_T(\lambda)$, $c_P(\lambda)$.
3. Calculation of the isolated thrust curve $T_{is}(V)$ corresponding to the flight condition with the given engine regime, $T_{is}(V)$ correction to the true $T(V)$ and effective thrust $T_{ef}(V)$.
4. Calculation of all flight performance with the free airplane aerodynamic characteristics corrected for the respective ground effect of each flight condition.
5. Set up of the contradictory pairs of the flight conditions and creation of Pareto set graphs.
6. Evaluation of optimal Pareto fronts on Pareto sets graphs as the optimization selection.

3.2. Aerodynamic Characteristics of the Aeroplane

A model study of a small two-seat sport airplane with a requirement for a short take-off and landing (STOL) was chosen. The airplane is designed as the high-wing arrangement in which 13 m^2 trapezoidal wing of the aspect ratio equals to 7.2. The wing is equipped with a combined high-lift device on the leading and trailing edge—a slotted leading edge (slat) and Fowler flap. The whole wing leading edge is equipped with slat; Fowler flap is installed on 60% of the wing trailing edge. Wing airfoil was developed from GA(W)-1 in order to increase lift coefficient. The airplane is drawn up with the standard side-by-side seat arrangement and a fixed taildragger landing gear with fairing.

Lift and drag aerodynamic characteristics were determined according to the methodology described in Appendix B for the following flight lift device configuration:

1. Take-off lift configuration—slat + Fowler flap 15°.
2. Cruise configuration—retracted lift devices.

The primary characteristics are determined without the propeller influence on the aerodynamic drag of the airplane and without the ground effect (see Figure 2). The characteristics correspond to moment-balanced states in the typical flight configuration. The primary lift and drag characteristics of take-off and landing configurations are corrected for the ground effect depending on the actual height above the ground. An additional airplane aerodynamic drag due to the propeller stream is included in the true propeller thrust.

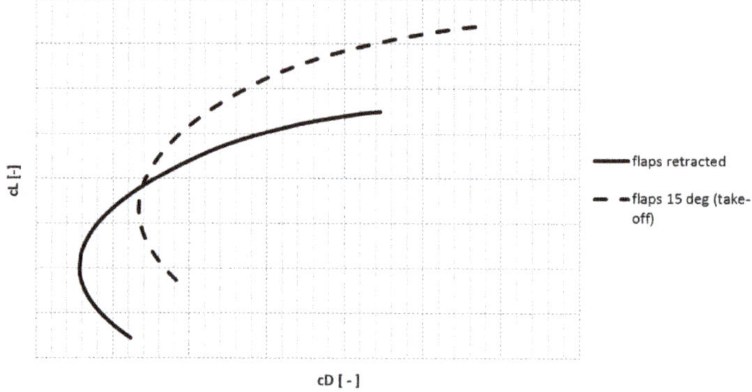

Figure 2. Polar graph of the aircraft for cruise (i.e., flaps retracted) and take-off configuration (i.e., flaps 15°).

3.3. Powerplant

The ROTAX 912 Aircraft Engine has been selected. The engine, primarily designed for this category of airplane, has take-off power of 73.5 kW and is equipped with a propeller speed reduction unit with gear ratio $i = 2.43$. The take-off regime corresponds to a maximum power at maximum permissible revolution per minute (RPM) for a short-term use, maximum 5 minutes (i.e., 5800 RPM at engine shaft); the continuous regime is 90% of the take-off power at reduced engine RPM (i.e., 5500 RPM at engine shaft) without any time limit; and the cruise regime (economic engine operation) means 75% of the continuous regime.

3.4. Propellers

3.4.1. Propeller Family

The propeller family is created as a group of geometrically similar propellers. Propeller blade geometry is derived from a three-blade propeller suitable for a 73.5 kW (100 hp) piston engine. The three-blade propeller was extended to other numbers of blades: two-blade, four-blade, five-blade, six-blade and eight-blade propeller with eleven pitch blade angles from 15° to 40° at 75% of radial distance. Geometric characteristics of the propeller are presented in Figure 3 (blade-chord distribution), Figure 4 (blade twist distribution) and Figure 5 (relative airfoil thickness). Clark Y airfoil is used. Range of propeller diameter of the total propeller family was from 1.1 m to 2.65 m, but each number of blades for a given pitch blade angle is only a part of this range. An extremely overloaded propeller that reaches the permissible engine RPM practically at the beginning of the take-off limits the smallest diameter. The maximum diameter represents a very lightly loaded propeller, and thus, the take-off is unacceptably extended, and the climbing rate is reduced.

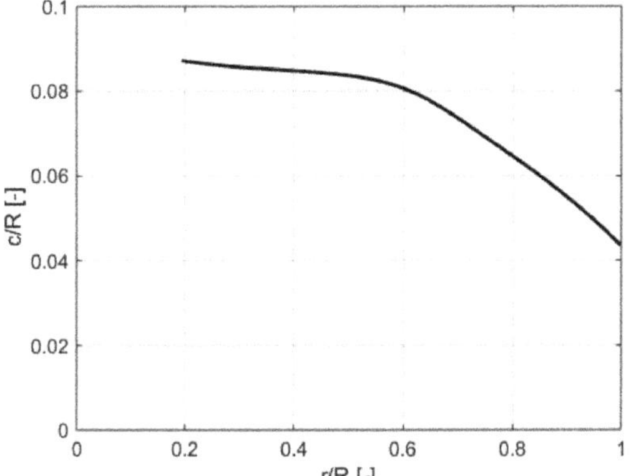

Figure 3. Propeller blade chord distribution.

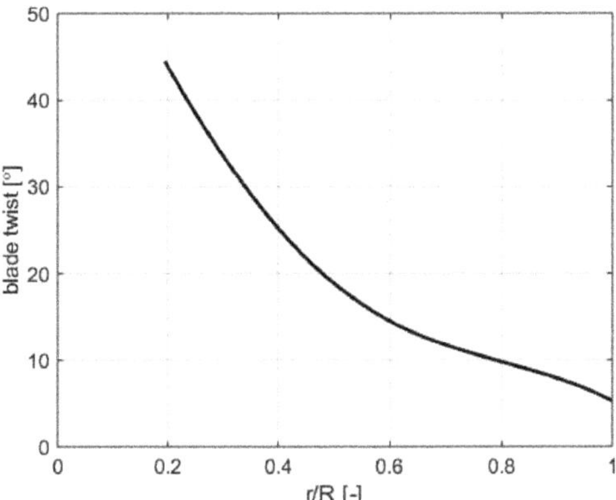

Figure 4. Propeller blade geometric twist distribution.

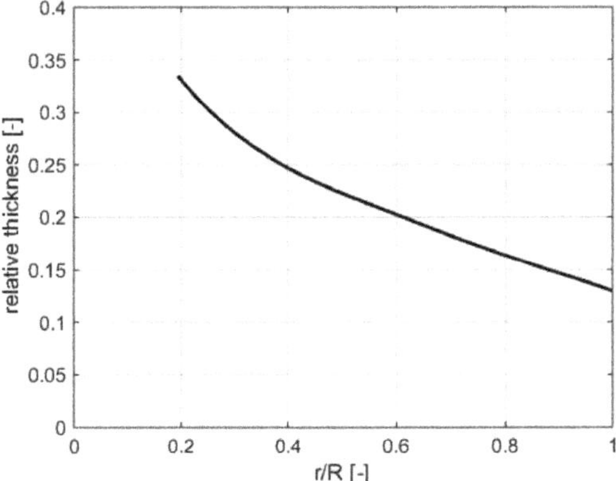

Figure 5. Propeller blade relative thickness distribution.

3.4.2. Aerodynamic Characteristics of Propellers

Aerodynamic characteristics of propellers needed to determine thrust curves are dependencies of the thrust and power coefficients on the advance ratio, i.e., $c_T(\lambda)$ and $c_P(\lambda)$. The vortex blade theory of an isolated propeller is used. Free helix vortex surfaces with a constant pitch leaving the boundary vortex of each blade generate a field of the induced velocities that adjust the magnitude and direction of free flow along the propeller blade. The aerodynamic forces acting along the blade can be considered as two-dimensional airfoil characteristics with the incoming free flow corrected to the induced angle of attack. The calculations of the propeller aerodynamic characteristics were performed using the numerical model [18,19]. Input geometric data involve, except diameter D and number of blades N, also the distribution of the chord length, twist and thickness of the airfoils used along the blade. The calculation was done for every blade pitch setting and number of blades. More detailed description of the method can be found in Appendix A.

3.5. Thrust Curves of the Propeller

The thrust curve of the power unit has the meaning of the thrust available in dependence on flight speed. First, the thrust curves of the isolated propellers were calculated, and then, these isolated thrusts were corrected for influence of the installation. The corrected thrust represents the true thrust. Because the additional drag of the airplane (both friction and pressure part) due to the increased speed of the propeller stream is thus dependent on the propeller thrust, it is preferable for the calculations of the flight performance to introduce so-called effective thrust. The effective thrust is the true thrust reduced by the additional propeller drag. The drag curve of the airplane without the additional propeller drag thus remains the same for all alternatives of the propeller propulsion units. The friction component depending on the thrust and wetted area influenced by the propeller flow was evaluated according to literature [20].

3.6. Flight Performance for Pareto Sets

Aircraft performance is computed by methods described in [6] and [21]. More detailed explanation can be found in Appendix B. Computation is performed for 0 meters international standard atmosphere (ISA), i.e., air pressure 101,325 Pa, air density 1.225 kg·m^{-3} and air temperature 288.15 K (15 °C). Pareto sets require the pairing of such requirements, which are contradictory. The respective pairs of flight conditions required to achieve their extreme values (minimum, maximum) act on each other so that increasing one flight condition towards the extreme decreases the second flight condition from its desired extreme.

In general, it can be expected that the propellers with good take-off performance are not good for reaching maximum flight speeds and vice versa. Therefore, for pairs of the flight conditions for Pareto set, the following are used:

1. Maximum horizontal flight speed (continuous engine regime)–Take-off distance (take-off engine regime).
2. Maximum horizontal flight speed (cruise engine regime)–Take-off distance (take-off engine regime).
3. Maximum horizontal flight speed (continuous engine regime)–Maximum rate of climb (take-off engine regime).
4. Maximum horizontal flight speed (cruise engine regime)–Maximum rate of climb (take-off engine regime).
5. Take-off distance (take-off engine regime)–Maximum rate of climb (take-off engine regime).
6. Maximum horizontal flight speed (continuous engine regime)–maximum horizontal flight speed (cruise regime).

4. Results

Pareto sets of six-selected flight condition pairs discussed in the Section 3.6 are plotted in Figures 6–11. Every point in the graph represents aircraft performance for one propeller defined by its number of blades, diameter and blade angle setting. Flight performance is computed according to the methodology described in Section 3. Marker types depend on the number of blades. For clarity, only limited areas near Pareto-optimal fronts from the all propeller family combinations are depicted. The broken lines connect points from Pareto-optimal fronts. Selection by means of speed power coefficient c_s is used for comparison (see c_s opt points in Figures 6–11). Performance with original constant-speed propeller is also shown for comparison. Results are further discussed in Section 5. Tables 1–6 contain coordinates of the points on the Pareto fronts for corresponding Pareto graphs. Pareto fronts are formed only by two-blade propellers; propeller diameter and blade angle are mentioned for every point. Performance with selected optimal propeller is also plotted in Figures 6–11).

Table 1. Points on the Pareto front for maximum horizontal speed (continuous engine regime) and take-off distance (see Figure 6).

$\varphi_{0.75}$ [°]	D [m]	Maximum Level Speed [km/h]	Take-Off Length [m]
22.5	1.95	214.54	125.25
20	2.05	212.22	122.61
20	2.0	204.68	119.63
17.5	2.15	204.37	119.03
17.5	2.1	197.7	115.99
15	2.25	192.88	115.05
17.5	2.05	190.97	114.55
15	2.2	187.06	112.34
15	2.15	181.18	111.1
15	2.1	175.25	110.63

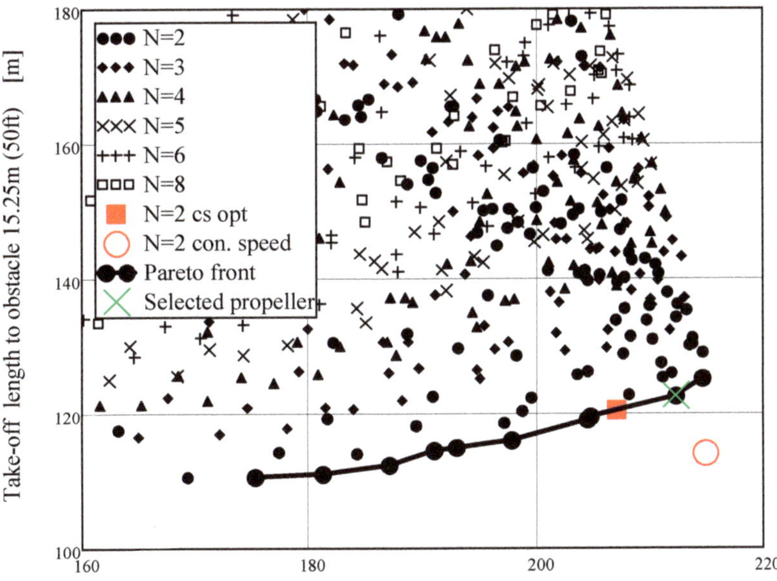

Figure 6. Pareto frontier for maximum horizontal speed (continuous engine regime) and take-off distance.

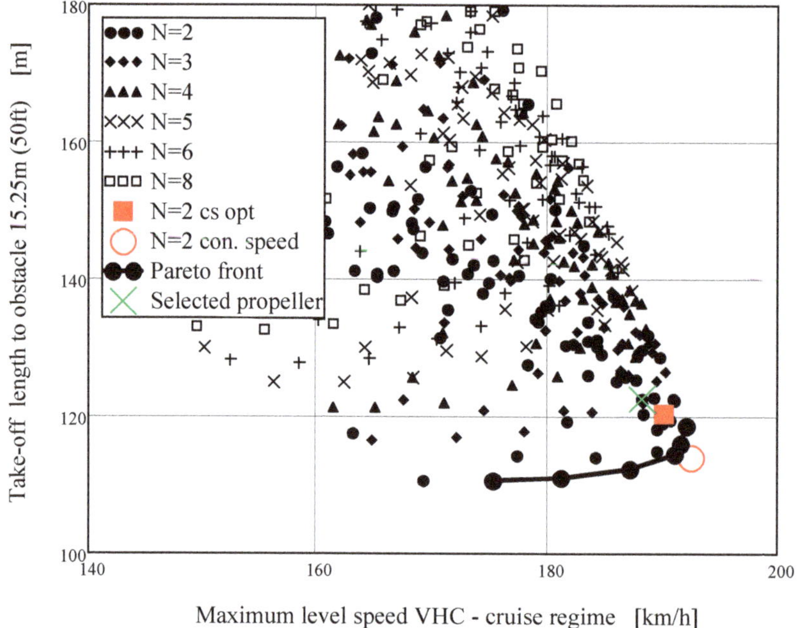

Figure 7. Pareto frontier for maximum horizontal flight speed (cruise engine regime)–take-off distance.

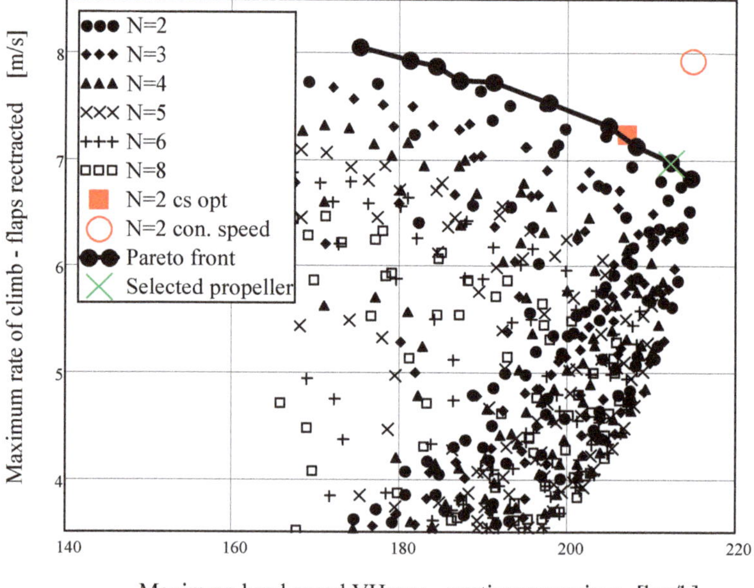

Figure 8. Pareto frontier for maximum horizontal flight speed (continuous engine regime) and maximum rate of climb (take-off engine regime).

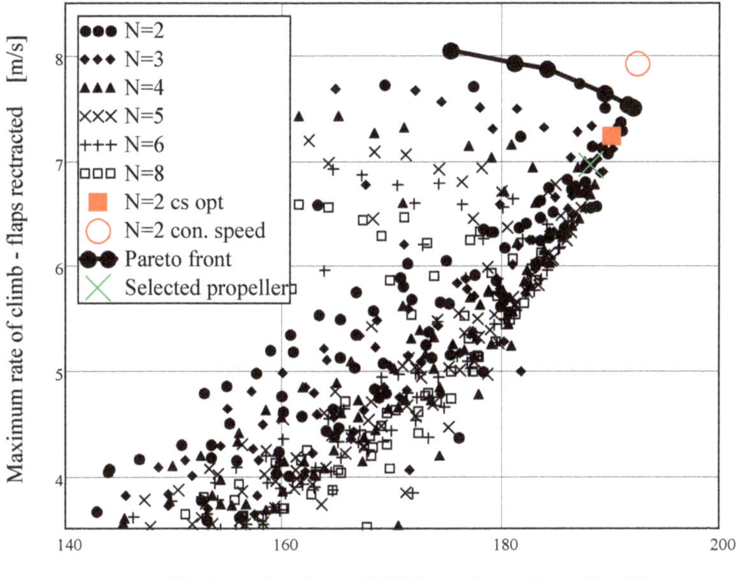

Figure 9. Pareto front for maximum horizontal flight speed, (cruise engine regime) and maximum rate of climb (take-off engine regime).

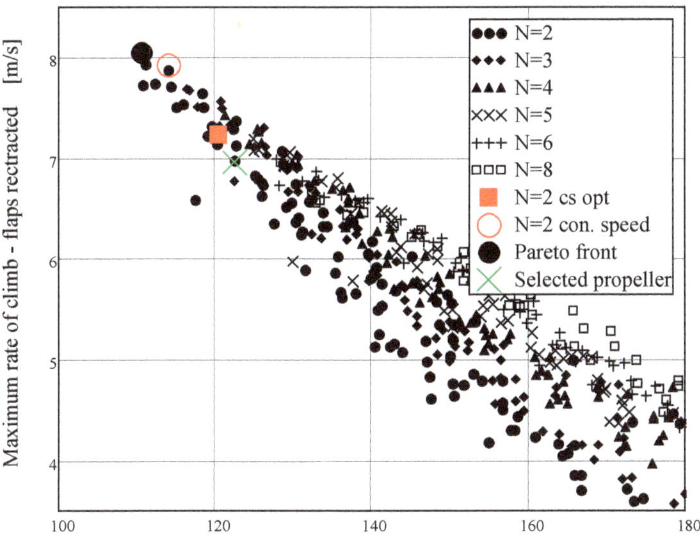

Figure 10. Pareto frontier for take-off distance (take-off engine regime) and maximum rate of climb (take-off engine regime).

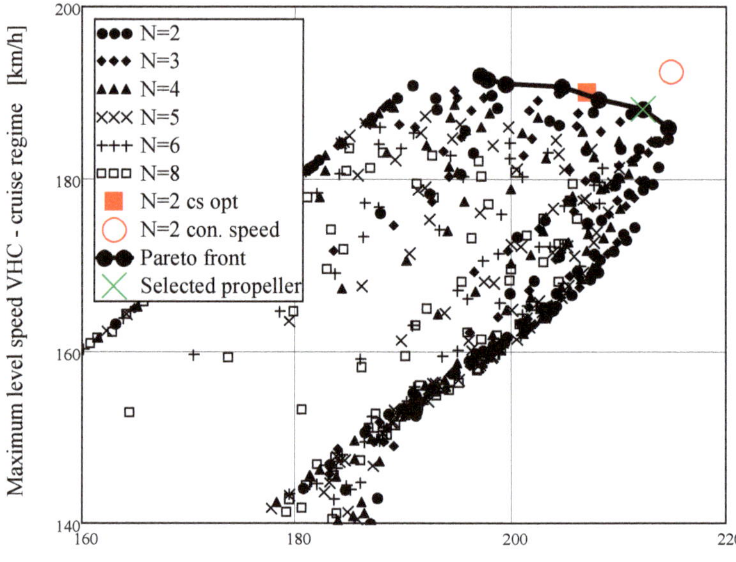

Figure 11. Pareto frontier for maximum horizontal flight speed (continuous engine regime) and maximum horizontal flight speed (cruise regime).

Table 2. Points on the Pareto front for maximum horizontal speed (continuous engine regime) and take-off length (see Figure 7).

$\varphi_{0.75}$ [°]	D [m]	Maximum Level Speed [km/h]	Take-Off Length [m]
20	1.95	194.4	118.59
17.5	2.1	191.53	115.99
17.5	2.05	190.97	114.55
15	2.2	187.06	112.34
15	2.15	181.18	111.11
15	2.1	175.25	110.63

Table 3. Points on the Pareto front for maximum horizontal speed (continuous engine regime) and maximum rate of climb (take-off engine regime) (see Figure 8).

$\varphi_{0.75}$ [°]	D [m]	Maximum Level Speed [km/h]	Maximum Rate of Climb [m/s]
22.5	1.95	214.54	6.823
20.0	2.05	212.22	6.973
22.5	1.9	207.97	7.122
20	2.0	204.68	7.315
17.5	2.1	197.70	7.531
17.5	2.05	190.97	7.723
15	2.2	187.06	7.735
17.5	2.0	184.18	7.876
15	2.15	181.18	7.926
15	2.1	175.25	8.049

Table 4. Points on the Pareto front for maximum horizontal speed (cruise engine regime) and maximum rate of climb (see Figure 9).

$\varphi_{0.75}$ [°]	D [m]	Maximum Level Speed [km/h]	Maximum Rate of Climb [m]
20	1.95	192.04	7.505
17.5	2.1	191.53	7.531
17.5	2.05	190.97	7.723
15	2.2	187.06	7.735
17.5	2.0	184.18	7.876
15	2.15	181.18	7.926
15	2.1	175.25	8.049

Table 5. Point on the Pareto front for take-off distance (take-off engine regime) and maximum rate of climb (take-off engine regime) (see Figure 10).

$\varphi_{0.75}$ [°]	D [m]	Take-Off Length [m]	Maximum Rate of Climb [m/s]
15	2.1	110.63	8.049

Table 6. Points on the Pareto front for maximum horizontal flight speed (continuous engine regime) and maximum horizontal flight speed (cruise regime) (see Figure 11).

$\varphi_{0.75}$ [°]	D [m]	Maximum Level Speed (Continuous Engine Regime) [km/h]	Maximum Level Speed (Cruise Regime) [km/h]
22.5	1.95	214.54	185.95
20	2.05	212.22	188.17
22.5	1.9	207.97	189.18
20	2.0	204.68	190.69
22.5	1.85	199.47	191.01
17.5	2.1	197.7	191.53
20	1.95	197.07	192.04

Figure 6 shows Pareto set for the Maximum level speed (continuous engine regime) vs. take-off length. Pareto front is located in the lower right-hand corner (short take-off and maximum speed). Pareto front is created uniquely by two-blade propellers. Two-blade propeller optimized by the means of the power speed coefficient c_s is on the Pareto-optimal set. Figure 7 shows similar situation for the cruise speed. In this case, Pareto-optimal set contains also uniquely two-blade propellers, but the solution obtained by the means of the speed power coefficient is no more on the Pareto-optimal set.

Figures 8 and 9 present Pareto-optimal front for the cases maximum level speed (continuous engine regime) vs. rate of climb and cruise speed vs. rate of climb. Both parameters should be maximal, i.e., optimal solutions are in the top right-hand part of the graph. Like before, optimal solution obtained by means of speed power coefficient is part of optimal set only for the case of the maximal continuous engine power.

Figure 10 presents Pareto frontier for the combination of take-off length vs. rate of climb. Optimal set is created only by single point in this case. Figure 11 presents Pareto frontier for the combination of maximum level speed (continuous regime) vs. cruise speed. Both parameters should be maximal, so the optimal set is located in the top right-hand part of the graph. As in previous cases, optimal set is represented only by two-blade propellers. Optimal solution obtained by means of the power speed coefficient is the part of this set.

5. Selection of Optimal Propeller

Optimal propeller is selected from the group of near to the Pareto fronts in the Figures 6–9 and Figure 11. The choice itself depends on the weight factors given to the different parameters, i.e., trade-off is always necessary. Take-off length was selected as the most important parameter due to the STOL category aircraft. Weight factors can be of course modified and tuned according to the designer requirements. The selected optimal propeller has two blades, diameter D = 2.05 m and pitch blade angle $\phi_{0.75}$ = 20°. This point is part of the Pareto fronts in Figures 6, 8 and 11. It is also close to the Pareto front in other cases. Aircraft performance with the optimal propeller is described in Table 7.

Table 7. Aircraft performance with chosen optimal propeller (diameter D = 2.05 m and pitch blade angle $\phi_{0.75}$ = 20°).

Maximum Level Speed (Continuous Engine Regime) [km/h]	Maximum Level Speed (Cruise Regime) [km/h]	Take-Off Length [m]	Maximum Rate of Climb [m/s]
212.22	188.17	122.61	6.973

6. Conclusions

Multi-objective optimization using Pareto sets of flight performance of 472.5 kg small STOL sport airplane powered by 73.5 kW (100 hp) engine to select a fixed propeller from a family of geometrical similar propellers is presented. The propeller family included propellers from two-blade to eight-blade with pitch blade angles from 15° to 40° and diameters range from 1.1 m to 2.65 m.

The optimization criteria required maximum level speed, maximum rate of climb and minimal take-off. The optimal Pareto fronts were investigated for four pairs of opposing flight conditions:

1. Maximum horizontal flight speed–Take-off distance.
2. Maximum horizontal flight speed–Maximum rate of climb.
3. Take-off distance–Maximum rate of climb.
4. Maximum horizontal flight speed–maximum horizontal flight speed (cruise regime).

Length of take-off path and maximum rate of climb relate to the take-off engine power regime; maximum horizontal flight speed was considered for both continuous and cruise power regime. Only four pairs of parameters are used for the optimization due to the fact that the last two pairs (take-off distance vs. rate of climb and cruise speed vs. maximal horizontal speed) in fact do not affect the choice of the optimal propeller.

Pareto-optimal study leads to the following conclusions:

1. Only two-blade propellers belong to all Pareto-optimal fronts.
2. One cut-off point of Pareto-optimal fronts for maximum horizontal speed corresponds to D = 1.95 m, pitch blade angle $\phi_{0.75}$ = 22.5° for the continuous engine regime and to D = 2 m, $\phi_{0.75}$ = 20° for cruise regime.
3. The opposite cut-off points of Pareto-optimal front for both maximum rate of climb and minimal take-off distance corresponds one propeller: D = 2.1 m, pitch blade angle $\phi_{0.75}$ = 15°.
4. The constant speed propeller achieves better performance than the best performance of the fixed pitch and ground adjustable propellers for all four optimization criteria.

Acceptable aircraft performance is reached for the propellers with two blades, diameter in the range from D = 1.95 to 2.2 m with a corresponding decrease of blade angle from $\phi_{0.75}$ = 22.5° to 12.5°. Optimal propeller is then selected from this group.

To compare Pareto multi-objective optimization with standard selection propeller from a family of geometrically similar propellers by means of the speed power coefficient, the design was performed for design speed from the middle of Pareto-optimal front of maximal horizontal speed for continuous engine regime. The single-mode optimization selection for two-blade propeller confirms Pareto optimization selection: diameter D = 2.05 m and pitch blade angle $\phi_{0.75}$ = 20°. The speed power coefficient methods are limited by single-regime design defined by given flight speed and engine regime, and the design is not clearly connected to flight performance. Two-blade propeller of this family considered as a constant speed propeller with diameter obtained by the speed power coefficient presents an increase in flight performance for the fixed pitch propeller.

Author Contributions: Author contributions are divided in the following way: conceptualization, S.S. and J.B.; methodology, J.K.; software, S.S. and J.K.; validation, S.S. and J.B.; formal analysis, S.S.; writing—original draft preparation, S.S. and J.K.; writing—review and editing, J.K.; visualization, S.S.; supervision, J.K. All authors have read and agreed to the published version of the manuscript.

Funding: This work was supported by the EU Operational Programme Research, Development and Education, under the Centre of Advanced Aerospace Technologies, project No. CZ.02.1.01/0.0/0.0/16_019/0000826, Faculty of Mechanical Engineering, Czech Technical University in Prague.

Conflicts of Interest: The authors declare no conflict of interest.

Nomenclature

D	Propeller diameter, Drag
H	Altitude
L	Lift
N	Number of propeller blades
P	Engine power
P_{max}	Maximum engine power
Q	Tangential force
S	Wing area
S_P	Propeller disc area
S_{wet}	Aircraft area wetted by the propeller flow
T	True propeller thrust
T_{eff}	Effective propeller thrust, $T_{eff} = (T - \Delta D)$
T_{is}	Isolated propeller thrust
T_{req}	Required thrust
U_0	Rotational flow speed on the propeller (freestream)
V	Flight speed
V_0	Axial flow speed on the propeller (freestream)
V_{Hmax}	Maximum horizontal flight speed–continuous engine regime
V_{HC}	Maximum horizontal flight speed–cruise engine regime
V_{min}	Stalling speed
V_1	Safe lift-off speed
V_{1P}	Flow velocity through the propeller disc (actuator disc)
V_2	Safe take-off climb speed (Safe speed of transition)
W	Total flow speed on the propeller
ΔD	Airplane drag induced by the propeller, $\Delta D = \Delta D_{fr} + \Delta D_{pr}$
ΔD_{fr}	Friction component of airplane-propeller drag
ΔD_{pr}	Pressure component of airplane-propeller drag
R	Propeller radius
c	Blade chord
c_P	Power coefficient of the isolated propeller, $c_P = P/(\rho\, n_s^3\, D^5)$
c_T	Thrust coefficient of the isolated propeller, $c_T = T_{is}/(\rho\, n_s^2\, D^4)$
c_s	Speed power coefficient of the isolated propeller, $c_s = [\rho\, V^5/(P\, n_s^2)]^{1/5}$
c_D	Drag coefficient
c_{Dmin}	Minimum drag coefficient
c_L	Lift coefficient
c_{Lmax}	Maximum lift coefficient
$c_L{}^\alpha$	Lift-curve slope
f	Rolling friction coefficient
h_{arc}	Height of the transition arc
h_W	Distance of the wing under the ground
i	Gear ratio of the engine speed reducer
m	Aircraft weight
n, n_{max}	Engine speed, Maximum engine speed
n_s	Propeller speed per second

n_y	Lift load factor
r	Radial coordinate
u	Induced tangential speed on the propeller
v	Induced axial speed on the propeller
v_y	Rate of climb
η	Propeller efficiency, $\eta = c_T\,\lambda/c_P$
λ	Propeller advance ratio, $\lambda = V/(n_s\,D)$
λ_W	Wing aspect ratio
ρ	Air density
$\varphi_{0.75}$	Propeller pitch blade angle at 75% of the radial distance
Ω	Propeller angular velocity

Abbreviations

ISA	International Standard Atmosphere
STOL	Short Take-off and Landing

Appendix A. Computation of Propeller Characteristics

Method for the computation of propeller characteristics is based on the combination of blade element theory and vortex theory. Dimensionless variables are used for the computation of isolated propeller performance. Standard formulations are used (see [19]). Flight velocity is expressed by advance ratio λ

$$\lambda = \frac{V}{n_s D}. \tag{A1}$$

Thrust is expressed by thrust coefficient c_T

$$c_T = \frac{T}{\varrho n_s^2 D^4}. \tag{A2}$$

Power coefficient c_P is defined in a following way

$$c_P = \frac{N}{\varrho n_s^3 D^5}. \tag{A3}$$

Radial coordinate r is replaced by

$$\bar{r} = \frac{r}{R}, \tag{A4}$$

Dimensionless velocities are defined in a following way

$$\bar{u} = \frac{u}{\Omega R}, \tag{A5}$$

and dimensionless circulation $\bar{\Gamma}$ is used in following form

$$\bar{\Gamma} = \frac{\Gamma}{4\pi\Omega R^2}. \tag{A6}$$

Thrust and power coefficients can be determined from [19]

$$c_T = \pi^3 \int_{\bar{r}_0}^{1} \bar{\Gamma}\left(\bar{U}_1 - \frac{c_D}{c_L}\bar{V}_1\right)d\bar{r}, \tag{A7}$$

$$c_P = \pi^4 \int_{\bar{r}_0}^{1} \bar{\Gamma}\left(\bar{V}_1 + \frac{c_D}{c_L}\bar{U}_1\right)\bar{r}\,d\bar{r}. \tag{A8}$$

Velocities on the propeller blade is shown in the Figure A1. U_1 and V_1 are defined by

$$\bar{U}_1 = \bar{U}_0 + \bar{u}, \tag{A9}$$

$$\bar{V}_1 = \bar{V}_0 + \bar{v}, \tag{A10}$$

where u and v represent induced velocities (during normal propeller operation, u is negative, and v is positive). Vortex theory is used for the determination of induced velocities. System of horseshoe vortices is used (see Figure A2). Induced velocities are computed by the way described by Okulov [22].

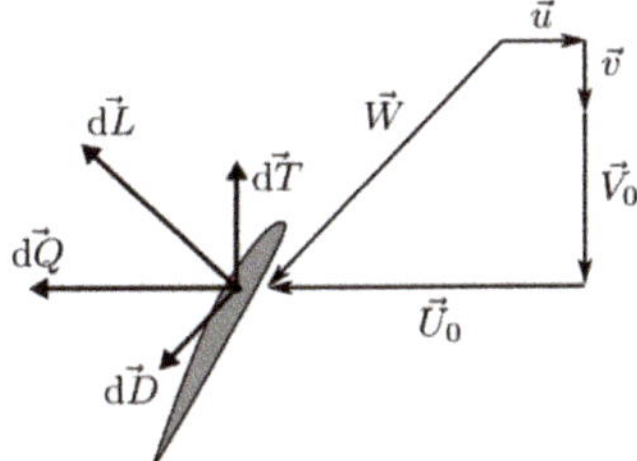

Figure A1. Flowfield and forces on the propeller blade.

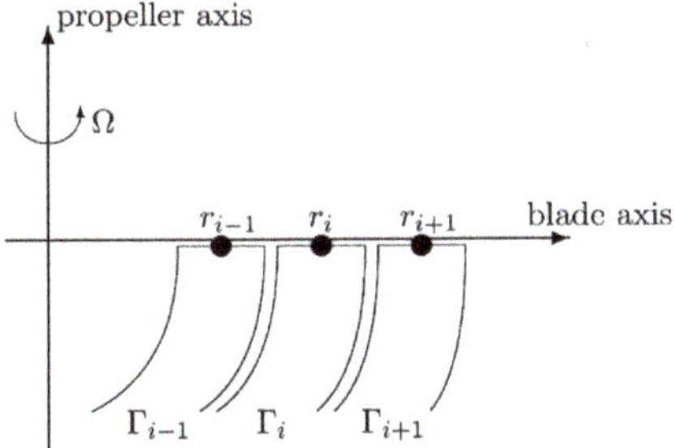

Figure A2. Illustration of the vortex system on the propeller blade.

Lift coefficient c_L is coupled with circulation by

$$\bar{\Gamma} = \frac{1}{2} c_L N \bar{c} \overline{W}_1. \tag{A11}$$

Airfoil section lift and drag coefficients are computed by means of the mathematical model described in [23]. Iterative procedure in MATLAB environment is used for the solution of the system of equations. The procedure was tested on the case of the three-blade propeller 5868-R6 described in [8]. Comparison of the computed results with experimental data is presented in Figures A3 and A4.

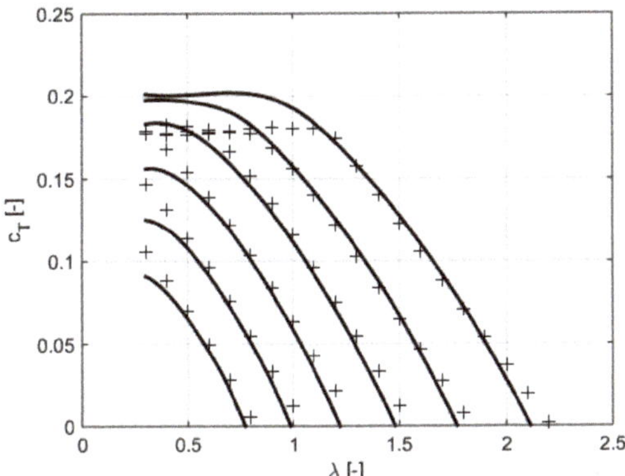

Figure A3. Comparison of computed values of thrust coefficient c_T (solid lines) with experimental results from [8] (discrete points) for the three-blade propeller 5868-R6.

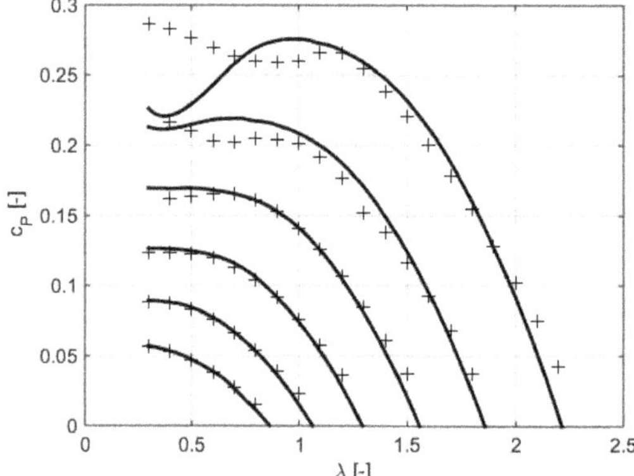

Figure A4. Comparison of computed values of power coefficient c_P (solid lines) with experimental results from [8] (discrete points) for the three-blade propeller 5868-R6.

Appendix B. Computation of Flight Performance

Appendix B.1. Thrust Curve of the Propeller Power Unit

The thrust curve of the power unit has the meaning of the available thrust in dependence on flight speed. First, the thrust curves of the isolated propellers are calculated, and then, the isolated thrust is corrected for influence of the installation. The corrected thrust represents the true thrust.

Because the additional drag of the airplane (both friction and pressure part) due to the increased speed of the propeller stream depends on the propeller thrust, it is preferable for the calculations of the flight performance to introduce so-called effective thrust. The effective thrust is the true thrust reduced by the additional propeller drag. The drag curve of the airplane without the additional propeller drag thus remains the same for all alternatives of the propeller propulsion units.

Appendix B.1.1. Thrust of Isolated Fixed-Pitch and Ground-Adjustable Propellers

Determination of the thrust curve of an isolated propeller requires, as the first step, the solution of the equilibrium propeller–engine rotational speed:

$$P(n) = \rho(H)n_s^3 D^5 c_P(\lambda). \tag{A12}$$

The solution determinates the root of the power equation expressed by the engine power curve $P(n)$ on one side and the propeller power described by the power coefficient $c_P(\lambda)$ for a given flight speed V on the other side. The equilibrium speed is solved for a number of speeds from the full range of the flight speed. The pair of the flight speed and the equilibrium rotational speed determinates the advance ratio λ and by help of the thrust coefficient $c_T(\lambda)$ the isolated thrust can be evaluated:

$$T_{is} = \rho(H)n_s^2 D^4 c_T(\lambda). \tag{A13}$$

The power engine curve $P(n)$ pertains to one of the three engine regimes: Take-off, Continuous and Cruise.

In the case that the equilibrium rotational speed with an increasing flight speed reaches the maximum allowed value n_{max}, it is assumed that the pilot will maintain this limit rotational speed by the throttle. The equilibrium rotational speed in such overload regime is thus the maximum speed n_{max}.

Appendix B.1.2. Thrust of Isolated Constant-Speed Propellers

As the constant-speed propeller keeps the rotational speed constant independently of the flight speed, then, the regulated rotational speed is actually the equilibrium speed. It is therefore possible to calculate the thrust directly by means of the thrust coefficient curve $c_T(\lambda)$ determined for the constant-speed propeller working in the required power regime.

Appendix B.1.3. Thrust Curve with Influence of Height

For calculating of thrust curves at different heights, the engine power of the piston engine at zero altitude can be converted to the given height H by a multiple factor k_N according to [24]

$$k_N(H) = (1 + 0.132)\frac{\rho(H)}{\rho_0} - 0.132 . \tag{A14}$$

The air density ρ at height H [m] corresponds to the standard dependence according to ISA:

$$\rho(H) = \rho_0\left(1 - \frac{H}{44308}\right)^{4.256} , \tag{A15}$$

where

$$\rho_0 = 1.225 \text{ kg/m}^3 , \tag{A16}$$

and the constant 44,308 m represents the ratio of two individual constants: the constant temperature decreases with high 0.0065 K/m and temperature of 288 K for $H = 0$ m.

The influence of the height on aerodynamic characteristics of propellers can be neglected.

Appendix B.1.4. True Thrust of Installed Propellers

The ratio of the true propeller thrust T and the thrust of isolated propeller T_i is estimated as the ratio of the actuator disc thrust of the propeller with a mean flow speed through the disc and thrust of the isolated actuator disc. The mean flow speed corresponds to the deceleration downstream due to a body behind the propeller. The mean value of the decelerated flow through the disc is determined from the equality of the flow momentum by the actuator disc with a constant and variable velocity.

The graphical dependence published in [19] is used to describe the change of the speed in the plane of the propeller disc caused by an engine nacelle. The graph shows the relative velocity drop $(\Delta V_{1P}/V_{1P})$ in the plane of the propeller disc in the area of the central part of the propeller. The central part is defined by the cross-section of the engine nacelle S_n. The velocity drop ratio depends on the cross-sectional area S_n and propeller disc area S_P and a compensatory analytical form of the graph has the form:

$$\frac{\Delta V_{1P}}{V_{1P}} = 0.2\frac{S_n}{S_P} - 0.08\sqrt{\frac{S_n}{S_P}} + 0.028 . \tag{A17}$$

The ratio of the thrusts is determined from the momentum of the induced velocity of the ideal actuator disc in the steady-state distance behind the disc and momentum of the ideal actuator disc with the mean speed. The ratio determinates the final correction factor to convert the isolated thrust to the true thrust:

$$k_{is} = \left(1 - \frac{S_n}{S_P}\left(\frac{\Delta V_{1P}}{V_{1P}}\right)\right)\frac{2V_{1P}\left(1 - \frac{S_n}{S_P}\left(\frac{\Delta V_{1P}}{V_{1P}}\right)\right) - V}{2V_{1P} - V} . \tag{A18}$$

The flow velocity V_{1P} through the propeller disc can be expressed depending on the thrust at flight speed V according to the theory of the ideal actuator disc:

$$V_{1P} = \frac{V}{2}\left(1 + \sqrt{1 + \frac{T_{is}}{\frac{1}{2}\rho(H)S_PV^2}}\right) . \tag{A19}$$

In the case of the single-engine airplane, the nacelle cross-section S_n is replied by the cross-section area of the engine part of the fuselage.

Appendix B.1.5. Effective Thrust

The effective thrust means the true propeller thrust reduced by the additional airplane drag generated by the propeller stream consisting of a friction and pressure component:

$$T_{eff} = T - \left(\Delta D_{fr} - \Delta D_{pr}\right). \tag{A20}$$

The friction component ΔD_{fr} is depended on the thrust and a wetted area of an airplane influenced by the propeller flow. Its mean value is presented in References [20,24]:

$$\Delta D_{fr} = 0.004\, T\frac{S_{wet}}{S_P}. \tag{A21}$$

The pressure component ΔD_{pr} is taken from Reference [19] for the engine nacelle case:

$$\Delta D_{pr} = \frac{\Delta V_{1P}}{V_{1P}} T.$$ (A22)

The conversion factor of the true trust to the effective thrust can thus be expressed in the form:

$$k_{eff} = 1 - 0.004\left(\frac{S_{wet}}{S_P}\right) - 0.2\frac{S_n}{S_P} - 0.08\sqrt{\frac{S_n}{S_P}} + 0.028.$$ (A23)

Appendix B.2. Curves of Available and Required Thrust and Power

Appendix B.2.1. Available Thrust Curve

The available thrust represents the effective thrust T_{eff}; see (A20). The available thrust curve is the dependence of the effective thrust on the flight speed V.

Appendix B.2.2. Required Thrust Curve

The required thrust is equal to the thrust that corresponds to the aerodynamic drag force without additional drag caused by propeller in a balanced horizontal flight at a steady flight speed V.

Calculation procedure for determining the required thrust:

Step 1: To determine the needed lift coefficient from the balance of lift and weight for the chosen flight speed V:

$$c_L = \frac{mg}{\frac{1}{2}\rho(H)V^2 c_L S}.$$ (A24)

Step 2: To state the corresponding drag coefficient from the aerodynamic polar of the respective flight configuration (without additional aerodynamic drag caused by propeller flow) for the corresponding lift coefficient. Drag coefficient c_D is determined from aerodynamic polar (Figure 2).

Step 3: The required thrust is equal to the aerodynamic drag:

$$T_{req} = \frac{1}{2}\rho(H)V^2 c_D S.$$ (A25)

Steps 1–3 shall be repeated from V_{min} (minimum horizontal flight speed at the given configuration–stalling speed) to the maximum flight speed V_{HC} (or V_{Hmax}) corresponding to the intersection of the required thrust curve $T_{req}(V)$ with the available thrust curve $T_{eff}(V)$ for a given altitude. An example of the curve of available and required thrust of a small sport aircraft at a cruise regime is depicted in Figure A5.

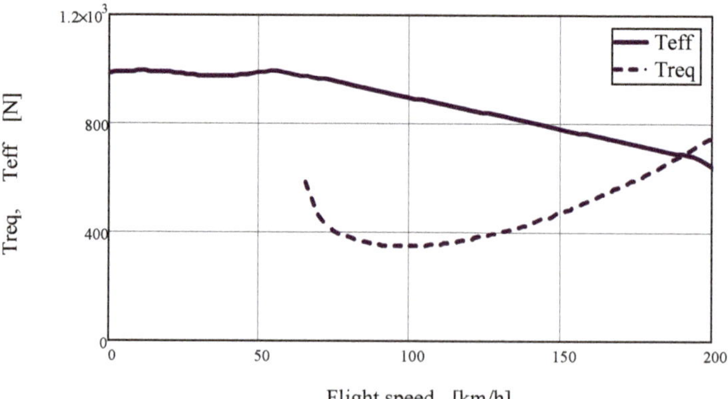

Figure A5. Curves of available and required thrust for a small sport aircraft $m = 473$ kg at a cruise configuration with three-blade propeller: $D = 1.75$ m, $\varphi_{0.75} = 22°$, maximum cruise-regime power 47.9 kW.

Stalling speed:

$$V_{min} = \sqrt{\frac{mg}{\frac{1}{2}\rho(H)V^2 c_{Lmax}S}} \,.$$ (A26)

Maximum flight speed: see (A31).

Appendix B.2.3. Required Thrust Curve with Influence of Height

1. Ground influence

Due to the ground proximity, aerodynamic drag decreases and lift increases. A simple correction according to Reference [21] is introduced for induced part of the drag and the lift-curve slope:

$$c_{Dground} = c_{Dmin} + (1 - \sigma)(c_D - c_{Dmin}) \,,$$ (A27)

where the correction factor σ is dependent on the height h_W of the wing above the ground, the aspect ratio of the wing λ_W and its area S:

$$\sigma(h_W, \lambda_W, S) = e^{\left[-4.22\left(\frac{h_W}{\sqrt{\lambda_W S}}\right)\right]^{0.768}} \,,$$ (A28)

$$c_{Lground} = \frac{c_L \pi \lambda_W}{\left[(1-\sigma)c_L^{\alpha}\right]} \,.$$ (A29)

2. Out of ground influence

Changes of aerodynamic forces are considered only by changing of the air density with height. The effect of Reynolds number changes on aerodynamic coefficients is neglected.

Appendix B.2.4. Available Power Curve

$$P_{eff}(V) = T_{eff}(V)\,V.$$ (A29)

Appendix B.2.5. Required Power Curve

$$P_{req}(V) = T_{req}(V)\,V.$$ (A30)

Appendix B.3. Maximum Horizontal Speed, Maximum Rate of Climb

Standard calculation models based on the curves of the required and available thrust and power are used to determine the maximum horizontal speed and maximum rate of climb; see, for example, Reference [20].

Maximum flight speed:

$$T_{req}(V) = T_{eff}(V) \rightarrow V = V_{HC}.$$ (A31)

Flight speed of maximum rate of climb V_{vymax}:

$$(P_{eff}(V) - P_{req}(V)) = \max \rightarrow V = V_{vymax} \,.$$ (A32)

Maximum rate of climb:

$$v_{ymax} = \frac{P_{eff}(V_{vymax}) - P_{req}(V_{vymax})}{mg} \,.$$ (A33)

Appendix B.4. Take-Off Distance

The take-off distance is understood to be the horizontal distance of the classic take-off procedure consisting of the ground round phase and three airborne phases: the ground flight, transition and climb out. The standard dynamic models of the take-off, discussed in, e.g., References [20,25], are adopted for the individual phases. The normal take-off considers the airplane with the take-off configuration of the lift device.

Appendix B.4.1. Take-Off Ground Run

The take–off ground run determines the distance to achieve the safe lift-off speed V_1.

Length of the ground run:

$$L_1 = \int_0^{V_1} \frac{V}{R_{ground\ run}(V)} dV \,,$$ (A34)

where safe lift-off speed V_1 is within the range (110–115)% of the stall speed with the ground effect. Stalling speed with the ground run effect corresponds to the relationship (A26):

$$V_{min\_ground\ run} = \sqrt{\frac{mg}{\frac{1}{2}\rho(H)V^2 c_{Lmax\_ground\ run} S}}\,, \tag{A35}$$

$$\text{for}: \quad c_{Lmax_{ground\ run}} = \frac{c_{Lmax}\pi\lambda_W}{\left[\left(1 - \sigma\left(h_{W_{ground\ run}}, \lambda_W, S\right)\right)c_L^\alpha\right]}. \tag{A36}$$

$R_{ground\ run}$ represents an accelerating force during the ground run:

$$R_{ground\ run}(V) = T_{eff}(V) - mgf - \left(c_{D_{ground\ run}} - c_{L_{ground\ run}}\,f\right)\frac{1}{2}\rho(H)V^2 S\,, \tag{A37}$$

with such a take-off angle of attack that satisfies the optimal condition:

$$\frac{dc_{L_{ground\ run}}}{dc_{D_{ground\ run}}} = \frac{1}{f}. \tag{A38}$$

Tangent to the lift-drag at the ground run point is equal to the inverse value of the rolling friction coefficient $1/f$.

Appendix B.4.2. Take-Off Ground Flight

The take–off ground flight corresponds to the acceleration phase of the ground level flight from the safe lift-off speed V_1 up to the safe take-off climb speed V_2.

Length of the ground flight:

$$L_2 = \int_{V_1}^{V_2} \frac{V}{R_{ground\ flight}(V)} dV, \tag{A39}$$

where safe take-off climb speed V_2 is required at least 120% of the stall speed in the ground flight regime. Stalling speed with the ground flight effect corresponds to the relationship (A26):

$$V_{min\_ground\ flight} = \sqrt{\frac{mg}{\frac{1}{2}\rho(H)V^2 c_{Lmax_{ground\ flight}} S}}\,, \tag{A40}$$

$$\text{for}: \quad c_{Lmax_{ground\ flight}} = \frac{c_{Lmax}\pi\lambda_W}{\left[\left(1 - \sigma\left(h_{W_{ground\ flight}}, \lambda_W, S\right)\right)c_L^\alpha\right]}. \tag{A41}$$

$R_{ground\ flight}$ represents an accelerating force during the ground flight:

$$R_{ground\ flight}(V) = T_{eff}(V) - c_{D_{ground\ flight}}(V)\frac{1}{2}\rho(H)V^2 S. \tag{A42}$$

The aerodynamic coefficients correspond to the point on the lift-drag polar of the ground-run configuration (corrected to the ground effect) at which the lift coefficient is equal:

$$c_{L_{ground\ flight}}(V) = \frac{mg}{\frac{1}{2}\rho(H)V^2 S}. \tag{A43}$$

Both of these coefficients thus depend on the ground flight speed V.

Appendix B.4.3. Take-Off Transition

The take–off transition phase means a curved flight from the safe take-off climb speed to the transition straight climb. The transition is taken as a flight close to a vertical circular arc with a constant speed equal to the safe take-off climb speed V_2.

Horizontal length of transition arc:

$$L_3 = r\sin(\theta)\,, \tag{A44}$$

where the radius of transition arc r deepens on the load factor n_y caused by the curvilinear flight:

$$r = \frac{V_2^2}{\left[g\left(n_y - \cos(\theta)\right)\right]} . \tag{A45}$$

The maximum achievable value of the load factor at the constant speed V_2 is

$$n_{ymax} = \left(\frac{V_2}{V_{min\_ground\ flight}}\right)^2 . \tag{A46}$$

This maximum value corresponds to the $c_{Lmax\ ground\ flight}$ and leads to the shortest length of the take–off transition phase, but it is on the edge of the sharp flow separation. A smaller value n_y should be considered, e.g., 80% n_{ymax}.

The output angle from the transition arc is given by the difference of the available and required thrust at the transition speed:

$$\theta = arcsin\left[\frac{T_{eff}(V_2) - T_{req}(V_2)}{mg}\right], \tag{A47}$$

where the required thrust equals:

$$T_{req}(V_2) = c_{D_{ground\ flight}}(V_2)\frac{1}{2}\rho(H)V_2^2 S. \tag{A48}$$

As the ground effect during the transition flight is gradually decreasing, an approximate procedure based on a mean value height h_W of the wing above the ground is applied. The mean value can be estimated, e.g., as a mean value between the height of arc h_{arc} calculated with the ground effect of the take-off ground flight and without the ground effect (free flight).

$$h_{W_{transition}} = \frac{1}{2}\left[h_{arc}\left(\sigma\left(h_{W_{ground\ flight}}\right)\right) + h_{arc}\sigma(h_W \to \infty))\right], \tag{A49}$$

where the arc height is equal:

$$h_{arc} = r\left[1 - \cos\left(T_{eff}(V_2) - T_{req}(V_2)\right)\right]. \tag{A50}$$

Appendix B.4.4. Climb Out

The climb out is the phase of the steady climbing flight at the speed of the transition phase from the transition arc until the obstacle height h_{obs} (usually 15.25 m = 50 ft)) is reached:

$$L_4 = \frac{h_{obs} - \left(h_{W_{ground\ flight}} + h_{arc}\right)}{tg(\theta)}, \tag{A51}$$

where the arc height h_{arc} and take-off output angle θ are determined for $h_{W\_transition}$.

Appendix B.4.5. Total Take-Off Distance

Total take-off distance is equal to the sum:

$$L_{take-off} = L_1 + L_2 + L_3 + L_4. \tag{A52}$$

References

1. Betz, A. Schraubenpropeller mit geringstem Energieverlust. *Göttinger Nachrichten* **1919**, *1919*, 193–213.
2. Goldstein, S. On the Vortex Theory of Screw Propellers. *Proc. R. Soc. Lond. (A)* **1929**, *123*, 440.
3. Larrabee, E.E. Practical Design of Minimum Induced Loss Propellers. *SAE Trans.* **1979**, *88*, 2053–2062.
4. Adkins, C.N.; Liebeck, R.H. Design of Optimum Propellers. *J. Propul. Power* **1994**, *10*, 676–682. [CrossRef]
5. Hepperle, M. Inverse Aerodynamic Design Procedure for Propellers Having a Prescribed Chord-Length Distribution. *J. Aircr.* **2012**, *47*, 1867–1872. [CrossRef]
6. McCormick, B.W. *Aerodynamics, Aeronautics, and Flight Mechanics*, 2nd ed.; John Wiley & Sons: Hoboken, NJ, USA, 1995.

7. Hartman, E.P.; Biermann, D. *The Aerodynamic Characteristics of Full-scale Propellers Having 2, 3, and 4 Blades of ClarkY and R.A.F. 6 Airfoil Sections*; NACA Technical Report No. 640; Langley Memorial Aeronautical Laboratory: Hampton, VA, USA, 1938.

8. Hartman, E.P.; Biermann, D. *The Aerodynamic Characteristics of Six Full-scale Propellers Having Different Airfoil Sections*; NACA Technical Report No. 650; Langley Memorial Aeronautical Laboratory: Hampton, VA, USA, 1939.

9. Törnros, S.; Klerebrant, O.; Korkmaz, E.; Huuva, T. Propeller optimization for a single screw ship using BEM supported by cavitating CFD. In Proceedings of the Sixth International Symposium on Marine Propulsors SMP'19, Rome, Italy, 26–30 May 2019.

10. Mirjalili, S.; Lewis, A.; Mirjalili, S.A.M. Multi-objective Optimisation of Marine Propellers. *Procedia Comput. Sci.* **2015**, *51*, 2247–2256. [CrossRef]

11. MacPeill, R.; Verstraete, D.; Gong, A. Optimisation of Propellers for UAV Powertrains. In Proceedings of the 53rd AIAA/SAE/ASEE Joint Propulsion Conference, Atlanta, GA, USA, 10–12 July 2017.

12. Schatz, M.E.; Hermanutz, A.; Baier, H.J. Multi-criteria optimization of an aircraft propeller considering manufacturing. *Struct. Multidisc. Optim.* **2017**, *55*, 899–911. [CrossRef]

13. Dorfling, J.; Rokhsaz, K. Constrained and Unconstrained Propeller Blade Optimization. *J. Aircr.* **2015**, *52*, 1175–1188. [CrossRef]

14. Xie, G. Optimal Preliminary Propeller Design Based on Multi-Objective Optimization Approach. *Procedia Eng.* **2011**, *16*, 278–283. [CrossRef]

15. Ingraham, D.; Gray, J.; Lopes, L.V. Gradient-Based Propeller Optimization with Acoustic Constraints. In Proceedings of the AIAA Scitech 2019 Forum, San Diego, CA, USA, 7–11 January 2019.

16. Deb, K. *Multi-Objective Optimization Using Evolutionary Algorithms*, 1st ed.; Wiley: Hoboken, NJ, USA, 2001.

17. Tan, K.C.; Khor, E.F.; Lee, T.H. *Multiobjective Evolutionary Algorithms and Applications*, 1st ed.; Springer: Berlin/Heidelberg, Germany, 2005.

18. Klesa, J. Optimal Circulation Distribution on Propeller with the Influence of Viscosity. In Proceedings of the 32nd AIAA Applied Aerodynamics Conference, Atlanta, GA, USA, 16–20 June 2014.

19. Alexandrov, V.L. *Vozdushnye Vinty*, 1st ed.; Oborongiz: Moscow, USSR, 1951.

20. Ruijgrok, J.J. *Elements of Airplane Performance*, 1st ed.; Delft University Press: Delft, The Netherlands, 1990.

21. Lowry, J.T. *Performance of Light Aircraft*, 1st ed.; AIAA: Renton, VA, USA, 1999.

22. Okulov, V.L. On the Stability of Multiple Helical Vortices. *J. Fluid Mech.* **2004**, *521*, 319–342. [CrossRef]

23. Korkam, K.D.; Camba, J., III; Morris, P.M. *Aerodynamic Data Banks for Clark-Y, NACA 4-Digit and NACA 16-Series Airfoil Families*; NASA-CR-176883; Lewis Research Center, NASA: Cleveland, OH, USA, 1986.

24. Torenbeek, E. *Synthesis of Subsonic Airplane Design*, 1st ed.; Delf University Press: Delft, The Netherlands, 1982.

25. Ojha, S.K. *Flight Performance of Aircraft*, 1st ed.; AIAA: Reston, VA, USA, 1995.

Article

Effect of Hot-Wet Storage Aging on Mechanical Response of a Woven Thermoplastic Composite

Theofanis S. Plagianakos [1,†,*], Kirsa Muñoz [2,†], Diego Saenz-Castillo [3,†], Maria Mora Mendias [3,†], Miguel Jiménez [2,†] and Vasileios Prentzias [1,†]

[1] Hellenic Aerospace Industry S.A., GR 32009 Schimatari, Greece; prentzias.vasilios@haicorp.com
[2] Element Materials Technology Seville, C/Wilbur y Orville Wright 1, Aerópolis, 41300 San José de la Rinconada, Seville, Spain; kirsa.munoz@element.com (K.M.); miguel.jimenez@element.com (M.J.)
[3] FIDAMC, Foundation for the Research, Development and Application of Composite Materials, Avda Rita Levi Montalcini 29, Tecnogetafe, 28906 Getafe, Madrid, Spain; Diego.Saenz@fidamc.es (D.S.-C.); maria.mora.mendias@fidamc.es (M.M.M.)
* Correspondence: plagianakos.theofanis@haicorp.com; Tel.: +30-226-205-2247
† These authors contributed equally to this work.

Received: 27 November 2019; Accepted: 20 February 2020; Published: 24 February 2020

Abstract: The effect of hot-wet storage aging on the mechanical response of a carbon fiber polyether ether ketone (PEEK)-matrix woven composite has been studied. A wide range of static loads and selected cyclic load tests on the interlaminar fatigue strength were performed. Static tests were conducted in batch mode, including on- and off-axis tension, compression, flexure, interlaminar shear strength (ILSS) and fracture tests in Modes I, II and I/II. Respective mechanical properties have been determined, indicating a degrading effect of aging on strength-related properties. The measured response in general, as well as the variance quantified by batch-mode test execution, indicated the appropriateness of the applied standards on the material under consideration, especially in the case of fracture tests. The material properties presented in the current work may provide a useful basis towards preliminary design with PEEK-based woven thermoplastic composites during service in aerospace applications.

Keywords: carbon fibre thermoplastic composite; PEEK matrix; woven; aging; mechanical testing; static and fatigue

---

1. Introduction

Continuous-fiber reinforced thermoplastic composites are gaining attention in the aerospace industry for exhibiting advantages compared to thermoset composites, such as design and manufacturing flexibility, including multiple post-forming processes, and capability of being processed by a large range of traditional machining methods, fast cycle time and recyclability. As far as their mechanical performance is concerned, their enhanced impact resistance is a very attractive feature for selecting them in demanding lightweight applications. Moreover, in woven ply configurations they yield less anisotropic mechanical properties, which could be desirable in the context of conceptual design. In this context, the experimental determination of their mechanical properties over a wide range of static and fatigue mechanical tests, as well as quantification of the effect of aging on these properties, are extremely important for design and in-service monitoring purposes.

The main advantages of thermoplastic composites in comparison with thermoset composites have been very well described in the scientific literature to date. Excellent mechanical properties, good behaviour against impact, no need of cold storage owing to their long shelf-life and no chemical reaction during consolidation, which opens the possibility of short-time processing are among their most appealing features [1–4]. There is a significant amount of literature focusing on mechanical characterization of continuous-fiber thermoplastic composites. Chu et al. [5] characterized 3-D braided Graphite/polyether ether ketone (PEEK) composites by static tensile tests and experimentally determined time-dependent mechanical properties at various temperatures by performing relaxation experiments and dynamic mechanical analysis (DMA) tests. Hufenbach et al. [6] studied the strain-rate dependency of the mechanical properties of three thermoplastic composite materials, including impact energy absorption, at different temperatures, and additionally studied the effect of fiber–matrix interphase modification on transverse tensile strength. Liu et al. [7] developed a damage model accounting for fiber failure and matrix cracking and validated it with an open-hole compression test on woven PEEK specimens. Kuo et al. [8] performed 4-point bending tests on thermoplastic composites and studied the effect of molding temperature on flexure properties and failure modes. Boccardi et al. [9] used infrared tomography to study the effect of frequency on the temperature of cantilever glass- and jute-based woven thermoplastic composite specimens subjected to cyclic bending. Sorentino et al. [10] fabricated and characterized polyethylene naphthalate (PEN) thermoplastic composites at 100 °C by 3-point bending, DMA and impact tests. As far as shear loading characterization is concerned, Hufenbach et al. [11] performed Iosipescu tests on woven thermoplastic composites and used high-speed camera and digital image correlation for studying the deformation and failure in order to develop material modeling strategies in commercial finite element software. Zenasni and coworkers [12,13] experimentally studied the effect of fiber material and weave pattern on the fracture response of polyetherimide (PEI)-matrix-based woven thermoplastic composite specimens in Mode I and Mode II.

The effect of aging on static and dynamic response of engineering thermoplastics has been studied since the early 1990s. As a general conclusion, PEEK polymers (and, therefore, PEEK-reinforced composites) are not the thermoplastic polymers which may absorb the highest level of water. It has been reported by Buchman and Isayev [14] that PEEK composites may absorb approximately 0.2% of humidity, in comparison with other thermoplastic polymers, which may absorb above 1%. This is mainly caused by the semi-crystalline structure of PEEK, in comparison with other thermoplastic polymers which have a larger amorphous part. Béland [15] concluded that semicrystalline thermoplastics/carbon composites may absorb also around 0.2% of humidity while carbon/amorphous thermoplastic may absorb beyond 0.8%. The effect of thermal aging on carbon-fibre reinforced PEEK has been studied by Buggy and Carew [16], who performed static and fatigue flexure tests. Aging has been applied by two different methods: by storage for up to 76 weeks at high temperatures and by thermal cycling from room temperature (RT) to 250°C. They indicated a low sensitivity of the laminate flexural modulus and strength to ageing at 120 °C. Kim et al. [17] developed an analytical model for predicting the flexural properties degradation at high temperatures (540–640 °C) and performed 4-point bending tests for validation purposes. In the context of an implant application study, Schambron et al. [18] experimentally determined the effect of environmental ageing on static and cyclic flexure response of carbon-fibre/PEEK coupons and reported superior fatigue resistance compared to stainless steel. An experimental comparison of the bearing strength of woven-ply-reinforced thermoplastic or thermoset laminates at 120 °C after hygrothermal aging has been performed by Vieille et al. [19].

Thus, it can be concluded that the effect of aging on the mechanical response of woven thermoplastic composites has yet not been quantified over a wide range of mechanical tests. The present work aims to close this void by presenting an extensive test campaign performed to assess the mechanical properties of a high-performance woven carbon-fiber reinforced thermoplastic material in non-aged and aged conditions. Material characterization has been achieved by conducting mechanical tests according to (static tests) or based on (fatigue tests) ASTM and ISO standards, such as tension, compression, in-plane and interlaminar shear, flexure, Mode I, II and I/II fracture. Properties derived from static tests are reported in batch mode, to provide a measure of the test-type non-linearity and testing procedure repeatability. Moreover, the effect of aging is assessed by measuring mechanical properties after specimen environmental conditioning in hot-wet storage. Finally, albeit the main objective of the current work is to provide experimental data, in order to check the applicability of linear elastic fracture mechanics based on effective material properties, interlaminar fracture test cases in Mode I and II have been modelled in a commercial finite element software and the predictions were compared with measured values. The main conclusion of the present work is that aging leads to significant degradation of the strength in engineering woven thermoplastic composites, while stiffness-related properties seem to be rather insensitive to aging.

2. Materials and Methods

2.1. Specimen Manufacturing

Thermoplastic composite coupons for mechanical properties' characterization were cut from a set of carbon fiber/PEEK plates manufactured by FIDAMC (Getafe, Spain). The engineering polymer selected for this study consisted of Tenax®TPCL PEEK-4-40-HTA40 3K supplied by Toho-Tenax (Tokyo, Japan). The material was carbon-fiber-reinforced PEEK (CF/PEEK) fabric 5HS (5 harness satin), 0.31 mm nominal thickness per ply, 40% of resin weight fraction and a fiber areal weight (FAW) of 285 gsm. The stacking sequence of the thermoplastic laminates varied depending on the mechanical testing and defined by the standards, explained in detail in the next sub-sections. It should be noted that, due to the fact that the used CF/PEEK fabric was a 5HS (non-symmetrical), the 0° direction of each ply was defined as the warp direction of the roll material, as suggested by the material supplier.

On the basis of previous experience with other candidate manufacturing methods [20], the selected manufacturing process for the consolidation of the laminates was compression molding by hot-platen press. The utilized equipment is located at the FIDAMC facilities (Getafe, Spain). Each laminate was first-hand laid-up with the help of a manual welder and then located in a metallic frame which acted as material retainer. Two polyimide sheets with a release agent were placed at both sides of the laminate, acting as release films. Two metallic caul plates were also used in order to obtain a proper flat-surface finish. The consolidation cycle consisted of a heating ramp at approximately 2 °C/min up to a consolidation dwell of 30 min at 400 ± 10 °C with an applied pressure of 1 MPa.

2.2. Hot-Wet Storage Aging

During its service life, an aircraft is exposed to high temperatures and high levels of humidity. The properties of composite materials may be affected as a consequence of moisture absorption and high temperatures. A faithful replication of the environmental exposure during aircraft operation would require a cyclic conditioning procedure between hot/humid and cold/dry conditions, as dictated by relevant standards (MIL-STD-810 or other). In the proposed work, we have followed an accelerated conditioning procedure on the basis of common practice [21], which would form a small part of an extended experimental aging campaign towards material airworthiness certification.

In order to evaluate the degradation of mechanical properties, an environmentally conditioned testing scenario was considered according to ASTM D5229/D5229M. The selected method is a recommended pre-test conditioning method, consistent with the recommendations of the Composite Materials Handbook-17 [22]. Specifically, the procedure that has been followed in these test series is a conditioning procedure, BHEP, which covers non-ambient moisture conditioning of material coupons from the same batch as the ones tested, widely known as traveler specimens, in a humidity chamber (-H-) at a prescribed constant, conditioning the environment to equilibrium (-E-), periodic (-P-) coupon weighing being required. Two different types of test series were covered. On the one hand, room temperature (RT) tests were conducted on specimens without any previous conditioning. Room Temperature conditions are controlled to meet standard laboratory atmosphere conditions, which according to ASTM D5229/D5229M, are 23 ± 2 °C and 50% ± 10% relative humidity. On the other hand, parallel test series were performed on specimens that have been previously conditioned. These are known as Hot Wet (HW) tests.

The conditioning parameters are usually fixed according to the conditions to which aircraft structure may be subjected during its service life. This commonly means an equilibrium moisture weight in an 85% relative humidity environment and a temperature of 70 °C. Nevertheless, in order to achieve a lower completion time for the testing campaign, accelerated conditioning was carried out. In order to achieve a faster aging process, conditioning was performed at a higher temperature (80 ± 3 °C), under the same relative humidity (85% ± 5%). It was checked that glass transition temperatures for the assessed material, as provided by the manufacturer, were significantly higher than the accelerated conditioning temperature. By proceeding this way, a decrease in conditioning periods was achieved, speeding the rate of testing.

2.3. Experimental Methods

All mechanical tests presented in this work have been performed according to (static tests) or based on (fatigue tests) relevant ASTM or ISO standards. Each standard has been reported in the title of relevant the subsection of the following section. Static tests have been performed in a five specimen batch mode to provide a statistically valid representation of a material sample response under static loads. For the sake of cost savings in the case of such an extended test campaign, three specimens per test type have been tested in fatigue, each one at a load level representing a fatigue cycle regime: low cycle-(1.0×10^4 cycles), medium cycle-(1.0×10^5 cycles) and high cycle-fatigue (1.0×10^6 cycles). Each test was assumed to have concluded with a failure if either the specimen failed due to rupture or if a load-displacement slope decrease of at least 10% was detected compared to the slope at the 100th loading cycle. A stress ratio of $R = \sigma_{\text{min}}/\sigma_{\text{max}} = 0.1$ has been considered, which is typical for the characterization of carbon-fiber reinforced plastics (CFRP) under dynamic loading [23,24]. Regarding testing frequency, high frequencies may produce an increase in the specimen temperature, which leads to mechanical properties degradation [25]. To avoid this, a frequency of 5 Hz was applied in all fatigue tests and the temperature of the specimens was monitored by means of a thermocouple placed on each. The laboratories involved in the campaign considered a range of specimen temperatures from RT up to 35 °C to be acceptable. For the selected frequency of testing, no specimen temperature reached 35 °C at any of the tests performed, and thus no effect of frequency on the results was assumed. According to the testing strategy, the fatigue endurance limit was set to 1.0×10^6 cycles.

All tests of the non-aged composite system (RT) have been conducted in Hellenic Aerospace Industry (HAI) facilities on an INSTRON 8801 (Norwood, MA, USA) hydraulic testing machine at room temperature ($T = 25 \pm 1$ °C, 45% ± 5% humidity), equipped by an INSTRON load cell with a range up to 100 kN. In the case of fracture tests, a 5 kN Omega load cell has been used. In the case of double-cantilever beam (DCB) and mixed-mode bending (MMB) tests, a Philips SPC2050NC digital camera and Debut v4.08 by NCH software (Greenwood Village, CO, USA—non-commercial use edition) have been used to record crack propagation, whereas, in end-notch flexure (ENF) tests, crack propagation has been visually monitored. The camera-to-specimen distance was ca. 130 mm and focus has automatically been applied at a 640 × 80 pixel resolution and a rate of 30 frames per second.

HW tests have been performed on different Universal Testing Machines in Element facilities. Static tension (0°, ±45°, 90°) and compression (0°, 90°) tests were performed on a Zwick Z100 BS1, (Kennesaw, GA, USA) equipped with a Zwick load cell up to 100 kN. Static flexure, interlaminar shear strength (ILSS) and interlaminar fracture toughness tests were performed on a Zwick Retroline (INSTRON 5866) Universal Testing Machine, equipped with an INSTRON load cell having a range up to 10 kN. All static HW tests have been conducted under controlled temperature (70 °C) using thermocouples inside a temperature chamber (Thermcraft). Regarding fatigue, HW tests, tension (0°, ±45° laminates) and ILSS tests were performed on a Universal Dynamic Testing Machine INSTRON 8872, equipped with an INSTRON load cell up to 25 kN. For fatigue tension (90° laminate), a Universal Dynamic Testing Machine INSTRON 8801, equipped with an INSTRON load cell up to 100 kN, was used. In all fracture tests crack propagation has been visually monitored.

HW fatigue tests were performed at RT right after being environmentally conditioned. That choice was based on the assumption that there is an irreversible degradation of the material due to specimen conditioning. As reported in the literature [26], long-term environmental conditioning leads to irreversible changes that cause permanent property alterations within the matrix, the fiber surfaces and the fiber/matrix interface. Therefore, by testing unaged and aged specimens under fatigue loading at RT conditions, it is possible to evaluate the effect of permanent degradation caused by hygrothermal conditioning.

2.4. Finite Element Models

The numerical simulations for pure fracture Modes I and II have been performed using MSC MARC finite element (FE) software [27]. In both cases, cohesive elements have been integrated into the models in order to evaluate the prediction of delamination propagation with experimental results. The composite thermoplastic material is modelled using property values at ply level extracted from mechanical tests. The constitutive relation of the cohesive elements is based on Linear Elastic Fracture Mechanics (LEFM) and an exponential damage law is used [28]. The cohesive properties of the interface have been derived by experimentally determined values for normal traction (strength—Section 3.3.1), interlaminar shear traction (strength—Section 3.5.1) and critical energy release rates in Mode I (Section 3.5.3) and Mode II (Section 3.5.4). From these values, normal traction and fracture toughness in Mode I have been adapted according to the strategy formulated in [29] in order to yield a critical opening displacement allowing for solution convergence. As discussed in Sections 3.5.3 and 3.5.4, the modeling approach adopted herein, which has been successfully applied in unidirectional thermoset materials [28], failed to accurately estimate the maximum load in both fracture modes of the woven thermoplastic material.

In order to simplify the analysis of both DCB and ENF tests, 2D models have been created using plane strain full-integration elements (Type 11) for the bulk material and cohesive elements (Type 186) for the interface. There are four elements through the thickness for the DCB model and five elements for the ENF model because of the difference in specimen thickness. An element length of 0.75 mm has been selected for composite and interface in order to achieve an acceptable convergence rate.

3. Results and Discussion

The test campaign for material characterization in terms of mechanical properties includes static and fatigue tests. Static tests have been performed for the determination of basic mechanical properties to feed the structural design phase of thermoplastic parts, as well as for quantification of the degradation of these properties due to aging during service. Fatigue tests provide an overview of fatigue endurance of the material via measured S–N curves and validate the trend observed in the static tests concerning the effect of aging. Specifically, static tests include tension, compression, in-plane shear, flexure, interlaminar shear, and interlaminar fracture toughness in Mode I, II and I/II. Fatigue tests include interlaminar shear strength. The specimen nominal dimensions in fatigue were identical to the ones used in the corresponding static case. Wherever available, the measured data are compared with values provided by the manufacturer [30].

3.1. Specimen Dimensions and Lamination

The specimen dimensions have been selected in order to comply with the relevant standard. The lamination has been determined on the basis of relevant standards, while available experimental capabilities have also been taken into account, in order to remain within the measurable load range of available equipment. In Table 1, the geometry and lamination of the specimens are presented. The test type description is provided as T-0 (tension warp), T-90 (tension weft), C-0 (compression warp), C-90 (compression weft), FLEX (flexure), T-45 (tension ±45), ILSS (interlaminar shear strength), ILFT I (interlaminar fracture toughness Mode I), ILFT II (interlaminar fracture toughness Mode II), ILFT I/II (interlaminar fracture toughness Mixed-Mode).

Table 1. Specimens dimensions and lamination ($ denotes odd number of plies, as: FLEX—13 plies, ILSS—19 plies).

Specimen Data	T-0	T-90	C-0	C-90	FLEX	T-45	ILSS	ILFT I	ILFT II	ILFT I/II
Lamination	$[(0/90)]_3$	$[(0/90)]_6$	$[(0/90)]_{3S}$	$[(0/90)]_{3S}$	$[(0/90)_7]_S$	$[(45/-45)_8]_S$	$[(0/90)_{10}]_S$	$[(0/90)]_{5S}$	$[(0/90)]_{7S}$	$[(0/90)]_{7S}$
Length (mm)	250	175	79.4	79.4	154	250	40	125	160	155
Width (mm)	15	25	12.7	12.7	13	25	12	20	25	25
Thickness (mm)	0.9	1.9	1.9	1.9	4	4.8	6	3	4.2	4.2

3.2. Specimen Aging

For each test type, the respective moisture absorption curve, measured from corresponding traveler specimens (in-plane dimensions 25 × 25 mm), is reported. Figure 1 provides an overview of the conditioning process. On the basis of these curves, the moisture content uptake rate and water content at saturation during conditioning have been extracted in Table 2. For the determination of the slope, a measurement at saturation and the initial measurement have been used.

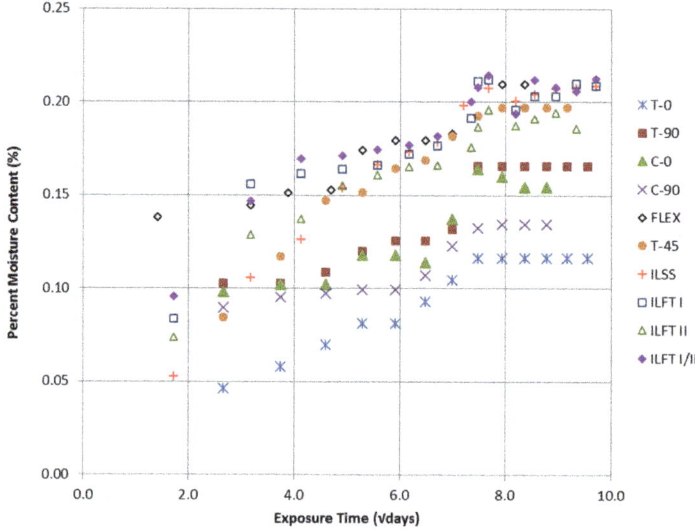

Figure 1. Conditioning curves for test types considered; the exposure time is expressed in square root of days.

Table 2. Moisture uptake parameters extracted from conditioning curves.

Moisture	T-0	T-90	C-0	C-90	FLEX	T-45	ILSS	ILFT I	ILFT II	ILFT I/II
Uptake Rate (%/days$^{0.5}$)	0.013	0.013	0.014	0.009	0.011	0.021	0.026	0.018	0.021	0.017
Maximum (%)	0.116	0.166	0.164	0.134	0.209	0.197	0.209	0.212	0.196	0.214

In order to comment on the trends observed in the conditioning process, the thickness of the relevant traveler specimen should be taken into account, since in-plane dimensions have been identical for all traveler specimens. It may be concluded that thicker traveler specimens, such as those considered for in-plane and interlaminar shear tests (including fracture toughness) yield generally higher moisture uptake rates and maximum values. This trend is attributed to the fact that moisture penetrates more easily across the sides of the specimens along thickness, which are uncoated compared to the upper and lower faces. Moreover, it is highly related to the effect of aging on the response in each test, as will be explained in the following sections.

Concerning comparison with data reported in the open literature, the maximum value of moisture uptake obtained for T-0 is very close to data reported by other researchers [31] (0.13%) and within the range set by previous studies on the moisture absorption of engineering thermoplastics [14,15].

3.3. Determination of Fiber-Dominated Properties

3.3.1. Static Tension of Woven Laminate Warp Direction (ASTM D3039)

The head velocity was 2 mm/min, as dictated by the standard. Strains were measured by strain gages: biaxial rosettes have been attached on each specimen, whereas an additional longitudinal strain gage has been attached on the opposite side of one specimen to ensure that bending remained below 10% throughout the test. Data for stress vs. longitudinal strain for RT and HW specimens are presented in Figure 2. It should be noted that the maximum load may differ from the value shown, as strain gages occasionally got detached at a lower load level. Moreover, in the legend of Figure 2 the custom naming convention pattern followed in all subsequent data figures is provided.

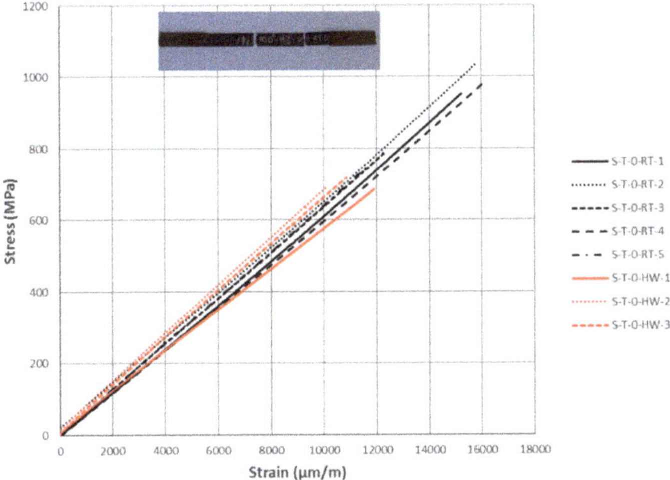

Figure 2. Static tension of warp-directional laminate: stress vs strain data and characteristic failure mode at RT. The description of tests follows a custom naming convention: S—static, T—tension, 0—warp direction, RT—room temperature test, HW—hot wet test, last digit—specimen number.

Elaboration of the above static tension test data yields the mechanical properties shown in Table 3. Properties are provided in terms of mean value (MV) and coefficient of variance (CV). CV is directly related to the standard deviation (STDEV) as: CV=STDEV/MV, and thus STDEV could be alternatively used for reporting the deviation intervals in the experimentally determined values.

The strain range between 1000 and 3000 με has been used for extraction of the tensile elastic modulus. The respective values provided in the manufacturer datasheet according to ISO 527-4 Type 3 are also listed for comparison.

Table 3. Effect of hot-wet storage aging on mechanical properties derived from static tension tests on warp-directional laminates.

Property	RT			HW	
	Mean Value	**CV (%)**	**Manufacturer Datasheet**	**Mean Value**	**CV (%)**
Tensile Strength (MPa)	889	13	955	701	2
Tensile Modulus (Gpa)	62.7	4	60	63.8	9
Poisson ratio	0.078	29	-	0.025	6

It may be observed that the elastic modulus is insensitive to aging (2% difference), whereas strength is strongly affected in a negative sense (21% reduction). This trend agrees well with results obtained by Solvay for PEEK thermoplastic polymer prepregs [32], and may be at least partially explained by the fact that the modulus is evaluated in the linear strain range (1000–3000 μm/m), while strength involves larger strains. It is reasonable to assume that in such a high demanding situation in terms of strain, the load-bearing capacity of the material appears to be sensitive to aging. As far as the Poisson ratio is concerned, it also seems to be affected, however, its measure is small and thus not expected to be of major importance as a design variable.

A typical failure mode is also illustrated in Figure 2. Lateral failure modes at multiple areas and various locations (LMV according to ASTM D3039) have been primarily observed. This trend has been expected, since the woven pattern of the fabric in the laminate prevents the occurrence of exploding modes. Similar failure patterns have been observed in RT and HW specimens, indicating that failure mode is rather insensitive to aging. Most failure patterns (8 out of 10) included failure near the tabs, indicating the redistribution of stresses along the specimen near/at failure load. The subsequent failure mode is related, partially at least, to inertia forces following the first failure, to a 3-D stress state involving local stress fields near the tabs, and to free-edge effects. The present approach focuses on the determination of the effect of hot-wet storage aging on the global mechanical properties of the woven thermoplastic material, while a detailed study of the failure mechanisms involved [11] would require special equipment, which has not been available in the context of this work.

3.3.2. Static Tension of Woven Laminate Weft Direction (ASTM D3039)

The selection of a different lamination compared to the warp-directional specimens, as listed in Table 1, is explained by the fact that the tests shown in this work are part of a larger test campaign where unidirectional laminates of various material systems are mechanically evaluated. Wherever possible, identical specimen dimensions for all material systems have been selected with the purpose of maximizing comparability for the same mechanical property evaluation. In this context, the tests on the weft-directional specimens have been similar to the warp-directional ones in terms of head velocity and strain gauge placement.

Stress vs. longitudinal strain data for RT and HW specimens are presented in Figure 3.

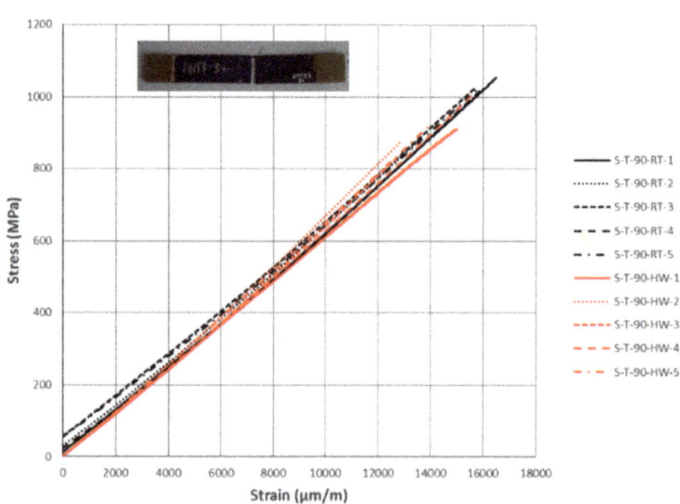

Figure 3. Static tension for weft-directional laminate: stress–strain data and characteristic failure mode.

Elaboration of the above static tension test data yields the mechanical properties shown in Table 4. The same analysis was applied as in the case of the warp-directional static tension tests. The respective values provided in the manufacturer datasheet according to ISO 527-4 type 3 are also listed for comparison.

Table 4. Effect of hot-wet storage aging on mechanical properties derived from static tension tests on weft-directional laminates.

Property	RT			HW	
	Mean Value	**CV (%)**	**Manufacturer Datasheet**	**Mean Value**	**CV (%)**
Tensile Strength (MPa)	973	6	909	907	8
Tensile Modulus (GPa)	56	2	60	61	2
Poisson ratio	0.039	16	-	0.039	-

Concerning the effect of aging, a similar trend is observed, as in the case of the warp-directional specimens. The elastic modulus appears to be slightly (9%) increased due to aging, while strength is slightly (7%) reduced. Such a trend for the modulus has been expected, as both laminations are similar, whereas the reduction in strength is less than expected.

A typical failure mode of an RT specimen is also illustrated in Figure 3. Lateral failure modes have been observed in all specimens, being less distributed along the specimen compared to the case of warp-directional laminates. Most specimens (7/10) failed in the vicinity of the tabs. These failures are attributed to pre-stress, due to excessive pressure on the tabs at the grips of the machine in order to avoid slippage during loading.

3.3.3. Static Compression of Woven Laminate Warp Direction (ASTM D695)

Two batches in total, each consisting of 5 RT and 4 HW specimens, were tested. The head velocity was 1.3 mm/min. Strains were measured by two unidirectional strain gages, each attached on the narrow part of each side of the dogbone specimen to ensure that bending remained below 10% up to buckling.

A typical load vs. longitudinal strain measurement is presented in Figure 4. A typical failure mode is also shown. All specimens failed under intralaminar shear.

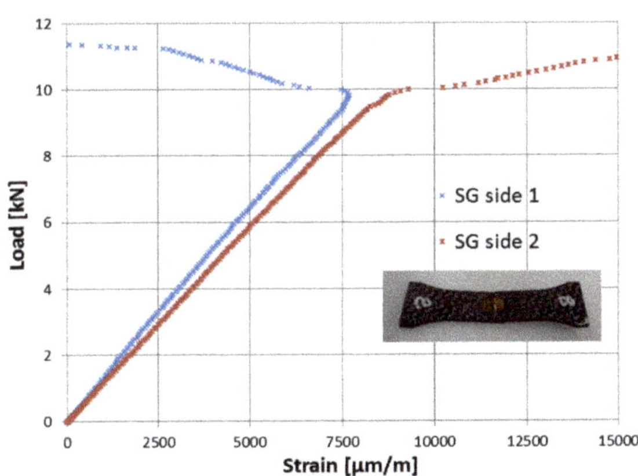

Figure 4. Compression test for warp-directional laminate: typical measured load vs longitudinal strain data and typical failure mode (SG: strain gauge).

Elaboration of the above static compression test data in the strain range between 1000 and 3000 με yields the elastic modulus shown in Table 5. The respective values provided in the manufacturer datasheet according to EN 2850 are also listed for comparison.

Table 5. Effect of hot-wet storage aging on compressive properties of warp-directional laminates.

Property	RT			HW	
	Mean Value (ASTM D695)	CV (%)	Manufacturer Datasheet (EN 2850)	Mean Value (ASTM D695)	CV (%)
Compressive Modulus (GPa)	55.0	3	59.2	60.3	2
Compressive Strength (MPa)	536	7	725	514	4

It may be observed that compressive modulus and strength exhibit low sensitivity (less than 10%) to aging. This trend agrees with the results reported in [32] for a similar material. The values measured for the modulus are close to those provided by the manufacturer, as they do not depend on the specimen geometry accounted for in the two standards. On the contrary, the higher thickness in most of the specimens due to the consideration of tabs in EN 2850 leads to higher strength values compared to the ones measured in the current study.

3.3.4. Static Compression of Woven Laminate Weft Direction (ASTM D695)

Two batches in total, each consisting of 5 RT and 4 HW specimens, were tested. The head velocity was the same as in the warp-directional case and strains were also measured accordingly. A typical load vs. longitudinal strain data curve is presented in Figure 5 along with a typical failure mode. All specimens failed under intralaminar shear.

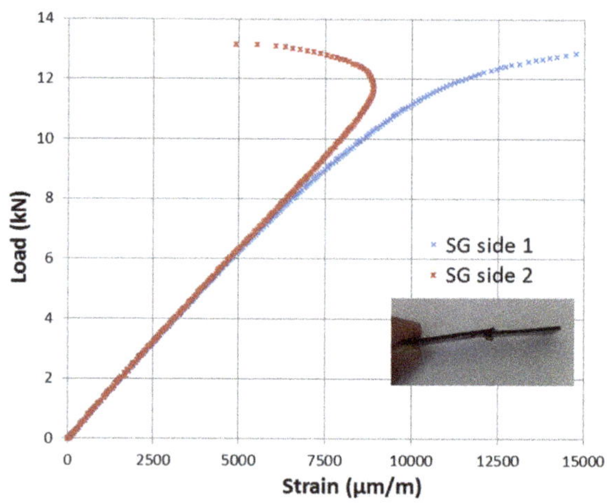

Figure 5. Compression test for weft-directional laminate: typical measured load vs. longitudinal strain data and typical failure mode.

Elaboration of the above static compression test data in the strain range between 1000 and 3000 $\mu\varepsilon$ yields the elastic modulus shown in Table 6.

Table 6. Effect of hot-wet storage aging on compressive properties of [90] laminates.

Property	RT			HW	
	Mean Value (ASTM D695)	CV (%)	Manufacturer Datasheet (EN 2850)	Mean Value (ASTM D695)	CV (%)
Compressive Modulus (GPa)	54.4	2	57.9	58.2	1
Compressive Strength (MPa)	541	4	712	504	9

As expected, the trends observed in the weft-directional lamination are similar to those reported in the warp-directional one (Table 5). Compressive modulus and strength exhibit low sensitivity (less than 10%) to hot-wet aging. The moduli measured are close to the value provided by the manufacturer, whereas there is a significant deviation in strength due to the application of different standards encompassing different specimen geometries.

3.3.5. Static Flexure (ASTM D7264)

Four-point bending tests have been performed in order to determine static flexural properties. Support span was 128 mm in order to comply with the span to thickness ratio suggested by the ASTM standard. The deflection was measured using an extensometer in a configuration similar to an LVDT. The head velocity was 1.5 mm/min.

The load vs. mid-span deflection for RT and HW specimens is presented in Figure 6, along with a characteristic failure mode.

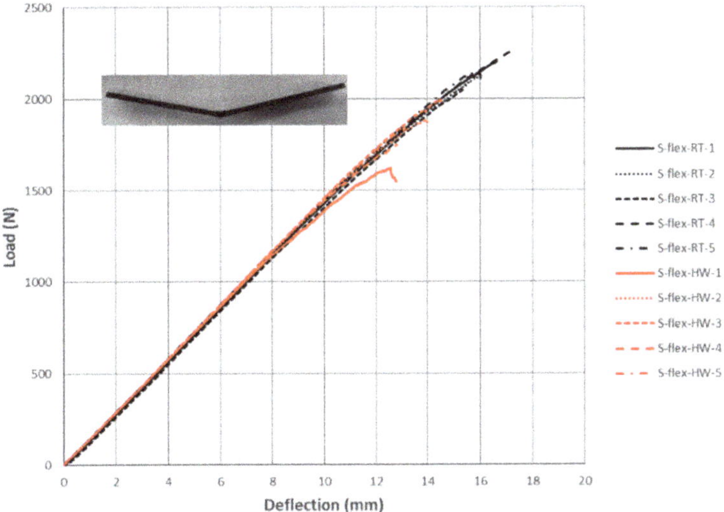

Figure 6. Static flexure: load-deflection data and typical failure mode.

Elaboration of the above test data yields the mechanical properties shown in Table 7. Respective values provided for a thinner lamination in the manufacturer datasheet according to EN 2562-A are also listed for comparison.

Table 7. Effect of hot-wet storage aging on the mechanical properties derived from static flexure tests.

Property	RT			HW	
	Mean Value	**CV (%)**	**Manufacturer Datasheet**	**Mean Value**	**CV (%)**
Flexural Strength (MPa)	974	4	1225	877	7
Flexural Modulus (GPa)	60.1	2	66.2	66.7	1
Failure strain (μm/m)	17337	4	-	13894	6

The elastic modulus appears to be slightly increased due to aging (10% increase), as in the case of tensile tests. On the other hand, aging leads to a drop of 16% in flexural strength and 20% in failure strain. Experimental studies reported in the literature [16] showed that thermal aging at 120 °C did not affect flexural modulus and strength. Therefore, the results presented herein indicate that it is the wet storage that mainly leads to strength degradation in a PEEK-matrix based composite configuration.

As far as failure modes of the specimens are concerned, all specimens failed under tension at the bottom surface, either at (5/10, failure description TAB according to ASTM D7264) or between loading noses (5/10, failure description TBB). According to the standard, these failure modes indicate valid flexural strength. This point is further supported by the acceptable variance in experimentally determined values for strength.

3.4. Determination of In-Plane Matrix-Dominated Properties

Static Tension of ±45 Woven Laminate (ASTM D3518)

In order to measure in-plane shear properties of the woven composite material, static tension tests on ±45 laminates have been performed. The head velocity was 2 mm/min, whereas strains were measured in the same manner as the other tensile tests presented above.

The stress vs. shear strain data for RT and HW specimens are presented in Figure 7, respectively. Each strain gage failed at approximately 15,000 με. The achieved range allowed for the evaluation of the 0.2% offset (yield) strength. Nevertheless, the load continued to rise after the failure of the strain gages and eventually reached a plateau at about 23 kN. Thus, the ultimate strength has been evaluated from the maximum load. At that stage, the plastic deformation of the specimens was visible with the naked eye, albeit no rupture of the specimen has occurred. That trend indicates a ductile response of the material, similar to that of metals, excluding failure in terms of breakage.

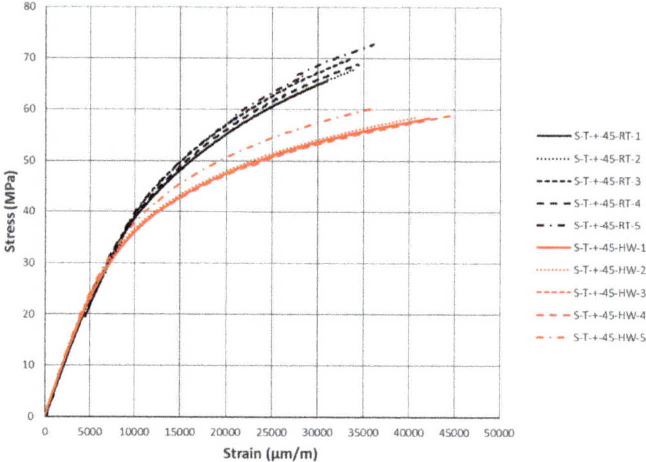

Figure 7. Effect of aging on stress–strain curves of in-plane shear strength (IPSS) test.

Elaboration of the above static tension test data yields the mechanical properties shown in Table 8. Respective values provided in the manufacturer datasheet (measured by using a proprietary testing method) are also listed for comparison.

Table 8. Effect of hot-wet storage aging on the mechanical properties derived from static tension tests on [±45] laminates.

Property	RT			HW	
	Mean Value	**CV (%)**	**Manufacturer Datasheet**	**Mean Value**	**CV (%)**
Offset Strength (MPa)	42.1	3	-	37.7	0.3
Shear Modulus (GPa)	4.32	4	5.03	4.46	2
Maximum Shear Stress (MPa)	98.4	3	-	82.9	7

As was also observed in the behavior of the tensile modulus in the previous tensile cases, the shear modulus, which is evaluated in a range of 2000 to 6000 µm/m, appears to be insignificantly affected by aging (3% increase). On the other hand, the tangential modulus beyond linear regime, as well as offset strength and maximum stress, are significantly lower in the case of the aged material (10% and 16%, respectively). Thus, the experimental results indicate that aging is more effective in larger strains occuring in the non-linear part of the response, which is prominent in that type of test. Considering that the maximum moisture absorption of in-plane shear specimens is slightly higher than in the tensile specimens, although not much higher, a reasonable phenomenological explanation of this trend is attributed to the thickness to width ratio of the specimens. In the in-plane shear case, the thickness to width ratio is approximately three times higher than in the tensile cases, and thus there is a wider distribution of moisture along the laminate plane, as moisture penetrates across the sides along thickness, which are not protected by a coating layer. As the strains increase, this trend gets more effective on the response and is enhanced by edge effects, which grow dominant. The experimental verification of that phenomenological explanation would be interesting, however, no appropriate equipment was available in the context of the present work to study the micromechanics involved during loading.

3.5. Determination of Out of Plane Matrix-Dominated Properties

3.5.1. Static Interlaminar Shear Strength Tests (ILSS) (ASTM D2344)

In the case of ILSS tests the support span was 24 mm, thus yielding a thickness aspect ratio of 4. The deflection was measured using the head displacement of the testing machine. The head velocity was 1 mm/min.

The load vs. mid-span deflection for RT and HW specimens is presented in Figure 8. A typical failure mode is also shown.

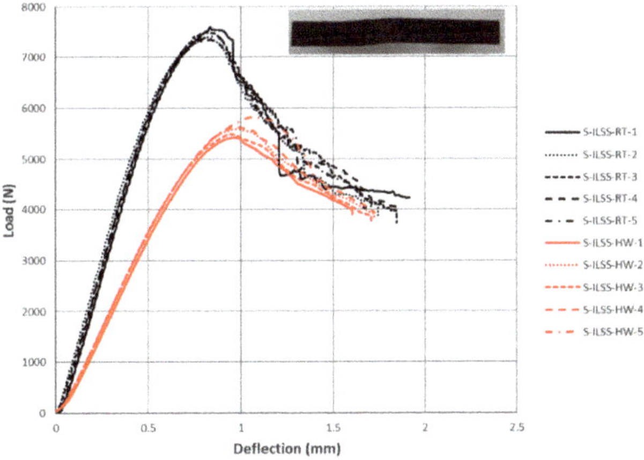

Figure 8. Effect of aging on load-deflection curves measured in ILSS test.

Elaboration of the above test data yields the mechanical properties shown in Table 9. For this test type, no data are included in the manufacturer datasheet.

Table 9. Effect of hot-wet storage aging on the mechanical properties derived from static ILSS tests.

Property	RT		HW	
	Mean Value	CV (%)	Mean Value	CV (%)
Maximum Load (N)	7453	1	5581	3
Short-beam Strength (MPa)	84.1	1	63.6	3

The interlaminar shear strength appears to be the most sensitive to aging among the properties studied so far, as a reduction of 25% has been experimentally determined. Moreover, the load-displacement slope is also significantly affected within both the linear and non-linear part of the response. This trend has been expected, since the ILSS test is a matrix-dominated test type. In addition, taking into account the dominance of the free-edge effects in such a low thickness aspect ratio (length/thickness) specimens, the ILSS experimental data led credibility to the phenomenological explanation provided in Section 3.4 concerning the effect of thickness to width ratio on the response. The thickness to width ratio of the ILSS specimens is 2.5 and 30 times higher than that of IPSS and tensile specimens, respectively, and thus they are more prone to a wider distribution of moisture in the specimen.

As far as the failure mode is concerned, the specimens failed under inelastic deformation and interlaminar shear. Each of these failure modes are considered acceptable in the standard, however, in the current case the failure appeared to include both modes as a result of the ductility of the matrix compared to thermoset composites.

3.5.2. Fatigue Interlaminar Shear Strength (ILSS) Tests

The effect of aging on interlaminar shear fatigue strength is presented in Figure 9. Aging has a clear degrading effect on the fatigue strength of the thermoplastic material, which is obvious in low-, medium- and high-cycle regimes. This trend is in-line with the experimental results of corresponding static tests and supports the argumentation based on the effectiveness of the thickness to width ratio on the enhanced sensitivity of ILSS specimens to hot-wet storage aging. It should be noted that all failures in fatigue have been attributed to a reduction in slope in the load-displacement loop beyond 10%. It should be also indicated that, as these tests are limited in quantity, statistical means cannot be drawn.

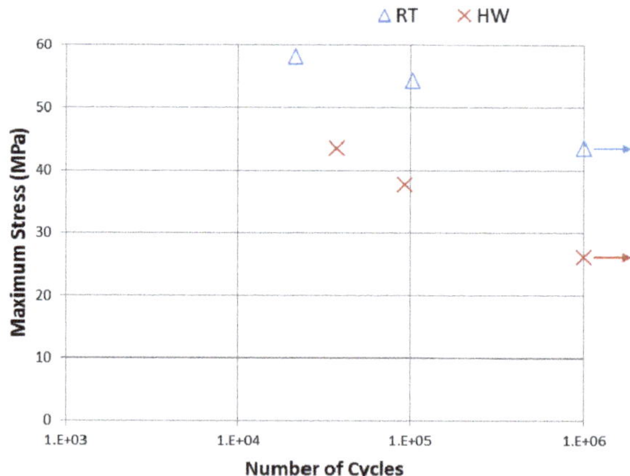

Figure 9. Effect of aging on fatigue interlaminar shear strength. Arrows to the right indicate that no failure has occurred.

3.5.3. Static Interlaminar Fracture Toughness Tests Mode I (ISO 15024)

According to the standard, an initial delamination of 60 mm has been built in by the intervention of a release film at the middle of the stacking sequence in the fabrication process. A head velocity of 1 mm/min has been applied. An initial loading has been considered for creating a crack propagation of 3–5 mm to eliminate fiber bridging effects, followed by a reloading stage for further crack propagation. In Figure 10a, measured load vs. opening displacement data for RT and HW specimens at the reloading stage are presented. A predicted instance of crack propagation is shown in Figure 10b. An interesting observation during the test was the "stick-slip" response during crack propagation, which resulted in the usability of few data points for the evaluation of fracture toughness. While in the case of unidirectional (UD) thermoset materials, a stable propagation has been measured [28], it being common practice and, as such, considered in the related standards, in the case of the woven thermoplastic material propagation occurred in maximum of four steps. As shown in Figure 10c, both the upper and lower face have been excessively deformed at each propagation step. Thus, strain energy was stored in the face up to a sudden release. At that point, the crack propagated to an arbitrary distance, indicating a highly non-linear response and leading to a relatively large variance in fracture toughness. This stick-slip response has not been observed in similar tests on PEI matrix-based woven thermoplastics reported in the open literature [13], whereas the values obtained for fracture toughness are in a similar range. Thus, the stick–slip response could be attributed to the type of matrix, adhesion between fibers and matrix and weave type.

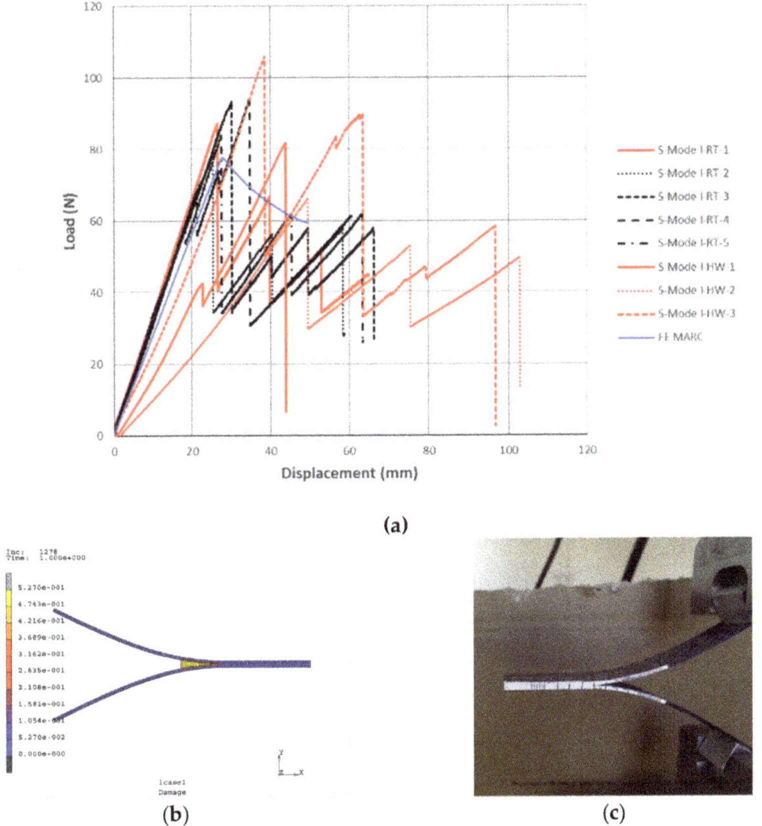

Figure 10. Mode I fracture test: (**a**) measured and predicted load vs. opening displacement data; (**b**) instance of crack propagation prediction; (**c**) typical deformed shape during test.

The FE model, which is based on LEFM, succeeds at predicting the slope of the load-displacement curves, whereas deviation is observed for maximum load. The deviation observed indicates that detailed unit-cell models and the application of appropriate failure criteria for woven thermoplastic materials should be applied in order to improve this prediction.

Elaboration of the measured data on the basis of Corrected Beam Theory (accounting for large-displacement correction) yields the interlaminar fracture toughness values presented as mean values in Table 10. For this test type no data are included in the manufacturer datasheet.

Table 10. Effect of hot-wet storage aging on interlaminar fracture toughness in Mode I.

Property	RT		HW	
	Mean Value	**CV (%)**	**Mean Value**	**CV (%)**
ILFT Mode I (J/m^2)	2490	12	3304	19

The aged specimens exhibited higher fracture toughness than the RT ones, which is not compatible with the trends observed in all other matrix-dominated test types, including the subsequent Mode II fracture tests. It could be partially related to plasticization effects at the crack front and the deformation mechanism of the faces. However, since the propagation evolves in limited steps, the contribution of each mechanism (strain stored in the faces within a geometrically non-linear state and purely fracture energy) can hardly be identified. For instance, the HW specimens may endure a larger deformation of the faces due to being more ductile as a result of conditioning process, which leads to the determination of higher G_{Ic} values.

3.5.4. Static Interlaminar Fracture Toughness Tests Mode II (ASTM D7905)

According to the standard, an initial delamination of 45 mm has been considered by applying a 0.012 mm release film at the middle of the lamination. A head velocity of 0.5 mm/min was applied. A non-precrack (NPC) and a subsequent precrack (PC) loading stage have been applied, each consisting of two tests for compliance calibration and an additional one for crack propagation. In Figure 11, measured load vs. transverse displacement data for RT and HW specimens in the crack propagation test of the reloading stage are presented, whereas measured and predicted instances of crack propagation are also shown. In contrast to the observations made in the DCB test, the woven thermoplastic material exhibits a stable response, similar to the one observed in the case of thermoset UD materials [28]. The FE model succeeds in accurately predicting the slope of the measured curve, whilst it fails to adequately estimate the measured maximum load for crack propagation. This shortcoming of the numerical model is largely attributed to the lack of implementation of appropriate failure criteria for woven thermoplastic materials encompassing interlaminar shear effects.

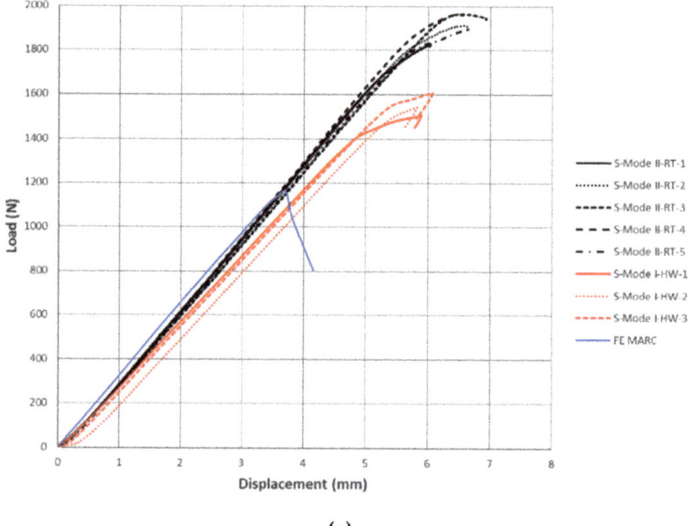

(a)

Figure 11. *Cont.*

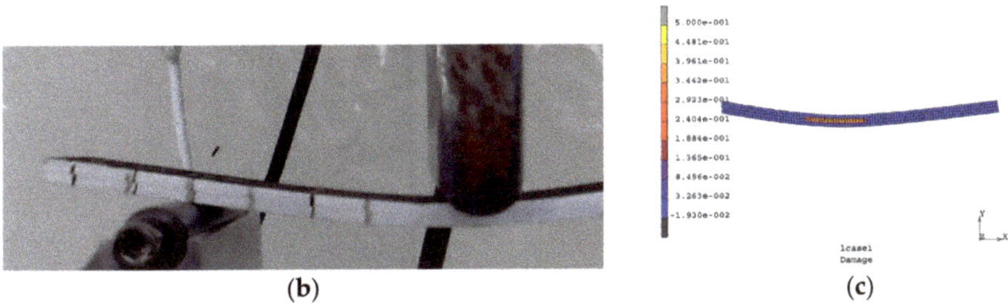

Figure 11. Mode II fracture test: (**a**) measured and predicted load vs. transverse displacement data; (**b**) typical deformed shape during test; (**c**) instance of crack propagation prediction.

Elaboration of the measured data yields the interlaminar fracture toughness values for RT and HW conditions, respectively, presented in Table 11. For this test type, no data are included in the manufacturer datasheet.

Table 11. Effect of hot-wet storage aging on interlaminar fracture toughness in Mode II.

Property	RT		HW	
	Mean Value	**CV (%)**	**Mean Value**	**CV (%)**
ILFT Mode II(J/m^2)	5900	8	4344	7

The aged specimens exhibited lower fracture toughness than the RT ones. Since interlaminar fracture toughness is a matrix-dominated mechanical property, that trend is in agreement with the behavior observed in other matrix-dominated tests, such as the IPSS and ILSS tests presented above. Nevertheless, the thickness to width ratio of the specimens involved is in the range of the IPSS ones, which is an additional argument for justifying the higher sensitivity to aging compared to tensile and flexure tests.

3.5.5. Static Interlaminar Fracture Toughness Tests Mode I/II (ASTM D6671)

A mode mixture of $R = G_{II}/(G_I + G_{II}) = 0.2$ has been selected for the RT specimens and a corresponding lever length of $c = 93$ mm. An initial delamination of 45 mm has been considered by applying a 0.012 mm release film at the middle of the lamination. A head velocity of 0.5 mm/min was applied. In Figure 12, measured load vs. opening displacement data for RT specimens are presented, along with instances of measured crack propagation. The "stick-slip" response during crack propagation has been also observed in this case, leading to relatively few data points and large variances. In the case of HW tests, the specimen failed at the root of the upper face at a maximum opening displacement between 3 and 10 mm. Thus, various mode mixtures ($R = 0.25$, 0.3 and 0.5) have been considered, albeit without success in eliminating that trend.

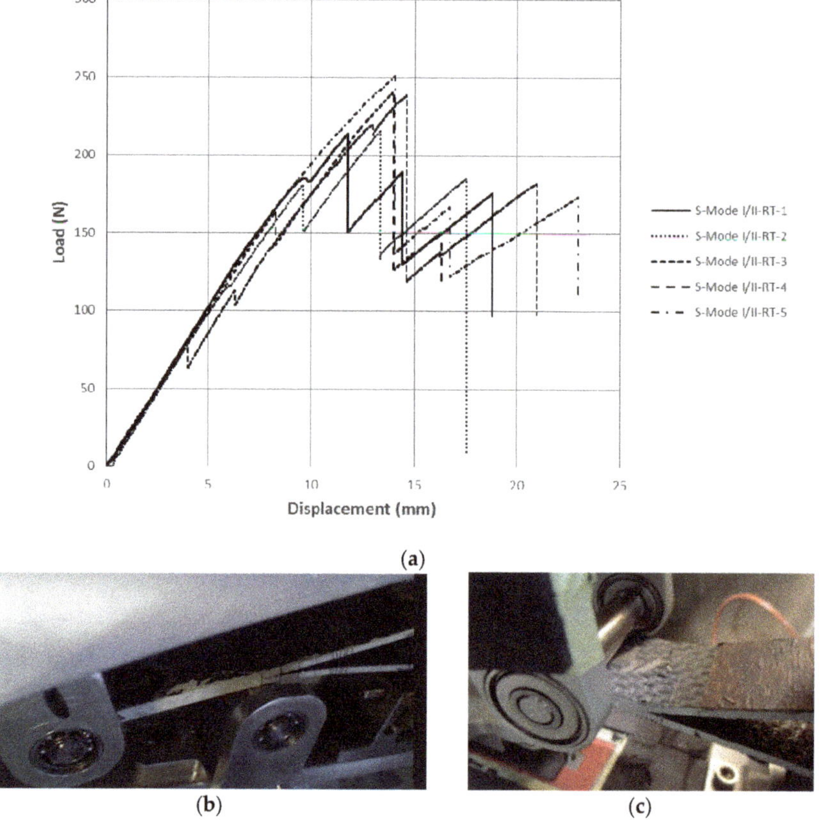

(a)

(b)

(c)

Figure 12. Mode II fracture test: (**a**) measured and predicted load vs opening displacement data RT specimens; (**b**) typical deformed shape during RT test; (**c**) failure mode at root of upper face during HW test.

Elaboration of the measured data yields the interlaminar fracture toughness values presented in Table 12. For this test type no data are included in the manufacturer datasheet.

Table 12. Effect of hot-wet storage aging on interlaminar fracture toughness in Mode I/II.

Property	RT (R = 0.2, 5 sp)		HW (R = 0.2, 1 sp)		HW (R = 0.25, 2 sp)		HW (R = 0.3, 2 sp)	
	Mean Value	CV (%)	Value	CV (%)	Mean Value	CV (%)	Mean Value	CV (%)
ILFT Mode I/II(J/m^2)	3050	31	521	-	814	6	796	14

The large variance observed indicates the highly non-linear nature of this test on woven thermoplastic PEEK materials. The aged specimens exhibited significantly lower fracture toughness than the RT ones. However, premature failure of the upper face, as well as the fact that the crack evolves in limited steps, did not allow for a proper estimation of fracture toughness in the case of the HW specimens. The latter two points might indicate that the standard should be further elaborated in order to prescribe a stable and uniform crack growth in woven thermoplastic laminates, especially under hot-wet aged conditions.

4. Conclusions

On the basis of the presented study, the following major conclusions may be drawn:

- In all test types, the hot-wet storage aging process led to moisture absorption comparable to values reported in the literature [31];
- Aging was found to significantly affect static properties related to the strength of the thermoplastic woven material in all tests considered, whereas stiffness properties were weakly sensitive to aging. This trend is in line with results reported in the literature for similar types of material [32];
- Matrix-dominated properties, such as interlaminar shear strength and mode II fracture toughness were found to be the most degraded due to aging. A reasonable justification for that, supported by the trends observed in the test campaign, is attributed to the high thickness to width ratio of specimens involved in corresponding tests. This ratio affects the distribution of moisture along the plane of the specimen and becomes dominant beyond the linear elastic part of the response;
- In the case of the DCB and MMB tests, a stick-slip response has been observed, which prevented stable crack propagation and led to relatively large variance in fracture toughness compared to UD thermoset laminates. In these cases, the contribution of each deformation mechanism (strain stored in the faces within a geometrically non-linear state and purely fracture energy) can hardly be identified;
- In the case of the MMB tests on HW specimens, premature failure occurred at the root of the crack upper face, indicating the limited applicability of the standard in such cases;
- Aging was found to have a clearly degrading effect on interlaminar shear fatigue strength. A justification for that trend is the same as in the corresponding static tests;
- Validation of the FE models developed, which were based on effective material properties to be used for initial design purposes, indicated the need to consider failure criteria specifically targeted to thermoplastic woven composite materials in order to accurately capture both load–displacement slope and maximum load in the case of fracture toughness test simulation.

Author Contributions: Contribution of each co-author to the current work may be distinguished as: Conceptualization, all; Investigation, K.M. and T.S.P.; Methodology, M.J., K.M., T.S.P., and V.P.; Resources, D.S.-C. and M.M.M.; Validation, M.J., K.M., T.S.P., and V.P.; Visualization, T.S.P.; Writing—original draft preparation, all; Writing—review and editing, T.S.P. All authors have read and agreed to the published version of the manuscript.

Funding: The current work has received funding from EU Horizon 2020 Clean Sky II project SHERLOC (Structural Health Monitoring, Manufacturing and Repair Technologies for Life Management of Composite Fuselage) under Grant Agreement No. CS2-AIR-GAM-2014-2015-01.

Acknowledgments: The authors from HAI would like to thank our ex-colleague Stavros Kalogeropoulos for his major assistance with the experimental work.

Conflicts of Interest: The authors declare no conflict of interest.

References

1. Biron, M. *Thermoplastics and Thermoplastic Composites*, 1st ed.; Elsevier: Oxford, UK, 2013.
2. Cogswell, F.N. *Thermoplastic Aromatic Polymer Composites*, 1st ed.; Butterworth-Heinemann Ltd.: Oxford, UK, 1992.
3. Vaidya, U.K.; Chawla, K.K. Processing of fibre reinforced thermoplastic composites. *Int. Mater. Rev.* **2008**, *53*, 185–218. [CrossRef]
4. Fujihara, K.; Huang, Z.-M.; Ramakrishna, S.; Hamada, H. Influence of processing conditions on bending property of continuous carbon fiber reinforced PEEK composites. *Compos. Sci. Technol.* **2004**, *64*, 2525–2534. [CrossRef]
5. Chu, J.N.; Ko, F.K.; Song, J.W. Time-dependent mechanical properties of 3-D braided graphite/PEEK composites. *SAMPE Q.* **1992**, *23*, 14–19.
6. Hufenbach, W.; Böhm, R.; Thieme, M.; Winkler, A.; Mäder, E.; Rausch, J.; Schade, M. Polypropylene/glass fibre 3D-textile reinforced composites for automotive applications. *Mater. Des.* **2011**, *32*, 1468–1476. [CrossRef]

7. Liu, H.; Li, S.; Tan, W.; Liu, J.; Chai, H.; Blackman, B.R.K.; Dear, J.P. Compressive failure of woven fabric reinforced thermoplastic composites with an open-hole: An experimental and numerical study. *Compos. Struct.* **2019**, *213*, 108–117. [CrossRef]

8. Kuo, W.S.; Fang, J. Processing and characterization of 3D woven and braided thermoplastic composites. *Compos. Sci. Technol.* **2000**, *60*, 643–656. [CrossRef]

9. Boccardi, S.; Carlomagno, G.M.; Meola, C.; Simeoli, G.; Ruso, P. Infrared tomography to evaluate thermoplastic composites under bending load. *Compos. Struct.* **2015**, *134*, 900–904. [CrossRef]

10. Sorentino, L.; De Vasconcellos, D.S.; D' Auria, M.; Tirillo, J.; Sarasini, F. Flexural and low-velocity impact characterization of thermoplastic composites based on PEN and high performance woven fabrics. *Polym. Compos.* **2018**, *39*, 2942–2951. [CrossRef]

11. Hufenbach, W.; Langkamp, A.; Hornig, A.; Zscheyge, M.; Bochynek, R. Analysis and modelling the 3D shear damage behaviour of hybrid yarn textile-reinforced thermoplastic composites. *Compos. Struct.* **2011**, *94*, 121–131. [CrossRef]

12. Zenasni, R.; Argüelles, A.; Viña, I.; Garcia, M.A.; Viña, J. Influence of weave type and reinforcement in the fracture behavior of woven fabric reinforced thermoplastic composites. *J. Mater. Sci.* **2005**, *40*, 2987–2989. [CrossRef]

13. Zenasni, R.; Bachir, A.S.; Argüelles, A.; Castrillo, M.A.; Viña, J. Fracture characterization of woven fabric reinforced thermoplastic composites. *J. Eng. Mater. Technol.* **2007**, *129*, 190–193. [CrossRef]

14. Buchman, A.; Isayev, A.I. Water absorption of some thermoplastic composites. *SAMPE J.* **1991**, *27*, 30–36.

15. Béland, S. *High Performance Thermoplastic Resins and Their Composites*; Noyes Publications: Park Ridge, NJ, USA, 1990.

16. Buggy, M.; Carew, A. The effect of thermal ageing on carbon fibre-reinforced polyetheretherketone (PEEK). *J. Mater. Sci.* **1994**, *29*, 1925–1929. [CrossRef]

17. Kim, J.; Lee, W.; Tsai, S.W. Modeling of mechanical property degradation by short-term aging at high temperatures. *Compos. Part B-Eng.* **2002**, *33*, 531–543. [CrossRef]

18. Schambron, T.; Lowe, A.; McGregor, H.V. Effects of environmental ageing on the static and cyclic bending properties of braided carbon fibre/PEEK bone plates. *Compos. Part B-Eng.* **2008**, *39*, 1216–1220. [CrossRef]

19. Vieille, B.; Aucher, J.; Taleb, L. Comparative study on the behavior of woven-ply reinforced thermoplastic or thermosetting laminates under severe environmental conditions. *Mater. Des.* **2012**, *35*, 707–719. [CrossRef]

20. Saenz-Castillo, D.; Martín, M.I.; Calvo, S.; Rodriguez-Lence, F.; Güemes, A. Effect of processing parameters and void content on mechanical properties and NDI of thermoplastic composites. *Compos. Part A-Appl. Sci. Manuf.* **2019**, *121*, 308–320. [CrossRef]

21. Vodicka, R. *Accelerated Environmental Testing of Composite Materials*; DSTO Aeronautical and Maritime Research Laboratory: Melbourne, Australia, 1998.

22. *Composite Materials Handbook CMH-17 Rev G Polymer Matrix Composites*; Volume 1-Guidelines, Chapter 2; USA Department of Defense: Aberdeen, WA, USA, 2011.

23. Michel, S.A.; Kieselbach, R.; Martens, H.J. Fatigue strength of carbon fibre composites up to gigacycle regime (gigacycle-composites). *Int. J. Fatigue* **2006**, *28*, 261–270. [CrossRef]

24. Barati, M.; Bahari-Sambran, F.; Saeedi, A.; Chirani, S.A.; Eslami-Farsani, E. Fatigue properties determination of carbon fiber reinforced epoxy composite by self-heating measurements. In Proceedings of the 24ème Congrès Francais de Mécanique, Brest, France, 26–30 August 2019.

25. Justo, J.; Marin, J.C.; París, F.; Cañas, J. The effect of frequency on fatigue behavior of graphite-epoxy composites. In Proceedings of the ECCM16-16th European Conference on Composite Materials, Seville, Spain, 22–26 June 2014.

26. Hammami, A.; Al-Ghuilani, N. Durability and environmental degradation of glass-vinylester composites. *Polym. Compos.* **2004**, *25*, 609–616. [CrossRef]

27. *MARC® 2012*; MSC Software Corporation: Santa Ana, CA, USA, 2012.

28. Plagianakos, T.S.; Muñoz, K.; Guillamet, G.; Prentzias, V.; Quintanas-Corominas, A.; Jimenez, M.; Karachalios, E. Assessment of CNT-doping and hot-wet storage aging effects on Mode I, II and I/II interlaminar fracture toughness of a UD Graphite/Epoxy material system. *Eng. Fract. Mech.* **2019**, *224*, 106761. [CrossRef]

29. Turon, A.; Davila, C.G.; Camanho, P.P.; Costa, J. An engineering solution for mesh size effects in the simulation of delamination using cohesive zone models. *Eng. Fract. Mech.* **2007**, *74*, 1665–1682. [CrossRef]

30. Tenax®-E TPCL-PEEK-HTA40 Data Sheet. Available online: https://docplayer.net/87745139-Product-data-sheet-tenax-e-tpclpeek-hta40.html (accessed on 9 January 2020).
31. Ma, C.C.M.; Yur, S.W. Environmental effects on the water absorption and mechanical properties of carbon fiber reinforced PPS and PEEK composites. Part II. *Polym. Eng. Sci.* **1991**, *31*, 34–39. [CrossRef]
32. Technical Data Sheet APC-2 PEEK Thermoplastic Polymer Prepreg, Solvay. Available online: www.solvay.us (accessed on 17 January 2020).

Article

Modular Multifunctional Composite Structure for CubeSat Applications: Preliminary Design and Structural Analysis

Giorgio Capovilla [1,*,†], **Enrico Cestino** [1,†], **Leonardo M. Reyneri** [2,†] and **Giulio Romeo** [1,†]

1 Department of Mechanical and Aerospace Engineering, Politecnico di Torino, 10129 Torino, Italy; enrico.cestino@polito.it (E.C.); giulio.romeo@polito.it (G.R.)
2 Department of Electronics and Telecommunications, Politecnico di Torino, 10129 Torino, Italy; leonardo.reyneri@polito.it
* Correspondence: giorgio.capovilla@polito.it
† These authors contributed equally to this work.

Received: 29 November 2019; Accepted: 21 February 2020; Published: 24 February 2020

Abstract: CubeSats usually adopt aluminum alloys for primary structures, and a number of studies exist on Carbon Fiber Reinforced Plastic (CFRP) primary structures. The internal volume of a spacecraft is usually occupied by battery arrays, reducing the volume available to the payload. In this paper, a CFRP structural/battery array configuration has been designed in order to integrate the electrical power system with the spacecraft bus primary structure. The configuration has been designed according to the modular design philosophy introduced in the AraMiS project. The structure fits on an external face of a 1U CubeSat. Its external side houses two solar cells and the opposite side houses power system circuitry. An innovative cellular structure concept has been adopted and a set of commercial LiPo batteries has been embedded between two CFRP panels and spaced out with CFRP ribs. Compatibility with launch mechanical loads and vibrations has been shown with a finite element analysis. The results suggest that, even with a low degree of structural integration applied to a composite structural battery, more volume and mass can be made available for the payload, with respect to traditional, functionally separated structures employing aluminum alloy. The low degree of integration is introduced to allow the use of relatively cheap and commercial-off-the-shelf components.

Keywords: CubeSat; CFRP; structural integration; functional integration; structural battery; embedded battery

---

1. Introduction

CubeSats, in recent years, have experienced a growing interest, and will be employed in support of larger spacecrafts or as the very own space segment. The original CubeSat standard was conceived in 1999 as a spacecraft with cubic shape and 100 mm sides (standardized today as 1 unit, 1U) and a mass up to 1 kg [1] (today up to 1.33 kg [2]).

The original objectives of the CubeSat project were mostly educational, although in recent years an increasing number of CubeSat missions has gone beyond the original objectives and is aiming at cutting edge technological and scientific objectives [3]. Currently, the majority of CubeSats operates in Low Earth Orbit (LEO) and the prevailing form factor is 3U [4] (the combination of three 1U units, with a maximum mass of 4 kg [2]).

Current missions rely on aluminum alloy primary structures, and composite materials are sometimes employed in secondary structures. Multifunctional integration is usually limited to a few components, e.g., solar cells mounted on external panels, and is not employed systematically [4].

On the other hand, space missions objectives and architectures are becoming increasingly complex, leading to sophisticated but overcrowded CubeSats [5]. Another issue has to do with the structural mass ratio, i.e., the ratio of structural mass and spacecraft total mass, which may not be always satisfactory. For example, the excellent STRaND-1 CubeSat has a structural mass ratio of 30% and the authors did not consider this figure satisfactory [6].

Although some nanosatellites standards employed structural integration systematically [7], with composite primary structures the integration can move forward, and a higher volume and mass can be allocated for the payload. The capacities of composite materials in terms of tailoring and embedding are greatly attractive, and are promising alternatives if the designer's objective is to obtain highly functionalized, lightweight structures. Composite primary structures for CubeSats, if properly designed, can produce lower stresses and displacements, and higher fundamental frequencies with respect to traditional and widely employed aluminum alloy structures [8].

In this paper, the design of a Carbon Fiber Reinforced Plastic (CFRP) integrated primary structure component is addressed. It is part of a 1U satellite bus and it integrates an embedded Electrical Power System (EPS), to demonstrate the possibilities of functionalized, advanced composite primary structures for pico and nanosatellites and in particular for CubeSats. The integrated EPS components are solar cells, batteries and the necessary electronic circuitry. The integration of the EPS with the CubeSat primary structure allows to reduce the overall mass and volume by eliminating non-energy-storing EPS components, e.g., the case and the electrodes, and, in addition, the CFRP laminates provide containment and protection for the batteries. The CFRP integrated structure is conceived to fit in a 1U commercial, aluminum alloy frame and occupies one external face of the cube.

The structure under study shares the design philosophy of the AraMiS project [9]. Its goal is to design, produce, and test a set of "smart tiles" (Table 1) that must be mounted on the six faces of the above-mentioned commercial frame (Figure 1a). The main requirement for the tiles is their modularity. As a result, a tile accommodates mainly one bus subsystem components (Figure 1b). The modularity allows to reduce design, assembly and testing time and costs. a further reduction of costs is obtained with the use of Commercial-Off-the-Shelf (COTS) components. Clearly, an adequate level of redundancy must be considered.

Table 1. Existing smart tiles.

Name	Code	Main Subsystem/Components	Reference
Power management tile	1B8	OBC (On Board Computer, simple microprocessor) EPS (solar cells, primary converter) AOCS (gyroscopes, magnetometers, magnetorquers)	Figure 1a, front
CFRP power management tile version 1	1B68_V1	EPS (solar cells, secondary batteries)	Section 2.2
CFRP power management tile version 2	1B68_V2	EPS (solar cells, secondary batteries) AOCS (magnetorquers)	Section 2.2
Telecommunication tile	1B9	TT&C components (frequency bands: 437 MHz and 2.4 GHz)	Figure 1a, bottom, Figure 1b, top and back-right
Reaction wheel tile	1B213A	AOCS (reaction wheel, gyroscopes, magnetometers) EPS (solar cells)	Figure 1b, front, bottom and back-left

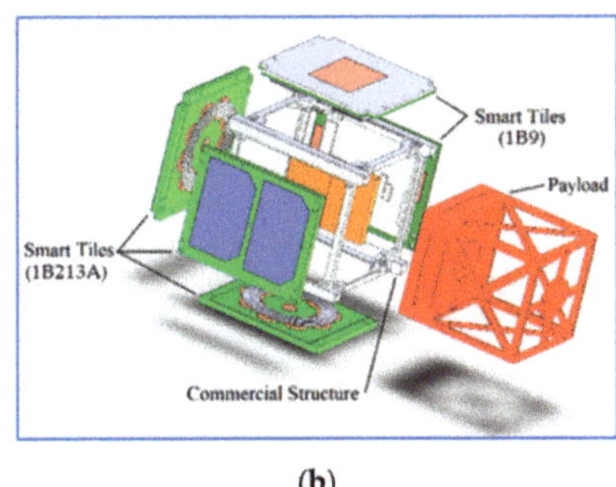

(a)   (b)

Figure 1. AraMiS CubeSat: (**a**) Prototype and (**b**) General architecture. Integrated bus functions include: power management (in the previous power management tile 1B8 and current work), Tracking, Telemetry and Command (TT & C, telecommunication tile 1B9), and attitude control by reaction wheels (reaction wheel tile 1B213A). Some functions like power generation are distributed over several tiles.

Existing smart tiles include the telecommunication tile, with Tracking, Telemetry, and Command (TT&C) subsystem components, the reaction wheel tile, with Attitude and Orbit Control System (AOCS) components and the power management tile. The aim of the present design is to improve the previous 1B8 power management tile. In that case, the primary structure function was given by Printed Circuit Boards (PCBs), which are not made with structural materials. Moreover, no energy storing function was implemented.

Besides space applications, energy storing composite structural components are studied in the automotive [10], aeronautical [11], and marine [12] sectors. Vehicle mass saving in this case is more useful than internal volume saving, with respect to the CubeSat case [13]. It has been shown that energy storing composites can significantly increase the range of aircrafts, as shown in [11], for piloted electric aircraft. Energy storing composites can also improve the performances of uninhabited, long-endurance aircrafts as High-Altitude Long-Endurance (HALE) vehicles exploiting solar energy or fuel cells [14,15].

2. Materials and Methods

The functionalization of the smart tiles must take into account a strong constraint, imposed by CubeSat deployment procedure. CubeSats are usually accommodated inside the launch vehicle in standard deployers, for example the Poly-Picosatellite Orbital Deployer (P-POD) for 1U-3U form factors [2]. The P-POD is an anodized aluminum prismatic box. A spring in the deployer pushes CubeSats into space, that are guided by their rails (the vertical rods shown in Figure 1a), in contact with rails on the deployer. For this reason, no component is allowed to protrude more than 6.5 mm normal to the plane of the rails (CubeSat mechanical requirements 3.2.3 and 3.2.3.1 [2]).

The functionalization of the primary structure has begun with the design of the reaction wheel tile prototype. Reaction wheels are attitude actuators that usually occupy CubeSats internal volume. The project consisted in merging the reaction wheel and the relative mechanisms and electronic components, the solar cells with related circuitry and the primary structure. The structure was still made with PCBs, made in epoxy resin and glass fibers. This material is usually addressed as FR-4, where FR stands for Flame Retardant.

The next step is represented by the design of an energy storing smart tile prototype, with the employment of composite materials. This will free internal volume, available to the payload, and will allow spacecraft bus mass saving.

2.1. Power Storage Composite Structures

There is a wide variety of integrated energy storage systems that can be conveniently described in terms of their degree of integration [11]. Traditional energy storing subsystems with no structural integration have zero degree of integration. An example of low degree of integration assembly is the embedded battery, where existing energy storing components are included in structural elements. Although the assembly has both functions, there is functional separation at the components level. At the other extreme, examples of high degree of integration assemblies are genuine structural batteries, whose components have the structural and the electrical function at the same time. One example of structural battery has been recently developed [10]. It consists of a unidirectional carbon fiber lamina where, in addition to the usual structural functions of the constituents, the fibers act as battery electrodes and the matrix acts as the battery electrolyte. As a result, the lamina can store electrical energy. Embedded battery systems come in a variety of configurations, the most usual ones are the laminate structure [16], the sandwich structure [12,17], and the stiffener structure [12]. The laminate structure consists of a classical composite material laminate with a set of batteries accommodated in the inner layers. The batteries are mechanically connected with the laminae thanks to a resin rich region that usually has an irregular shape. The sandwich structure houses the batteries in cavities of the core material, which usually has inferior mechanical properties with respect to the skins material. Finally, batteries can be placed in the inner regions of stiffeners.

Although high degree of integration batteries are promising and allow for greater mass and volume savings, to comply with the AraMiS project design philosophy, a low degree of integration has been chosen. Indeed, COTS components are relatively cheap and can easily be assembled and tested.

2.2. Power Management Tile Architecture

For the present design, an innovative configuration was chosen. It is a cellular configuration previously conceived for sailplanes wing boxes, in particular for the upper and lower skins [18] (Figure 2b). An evolution of this concept can as well be developed for other applications, as beam-like structures (Figure 2a) or innovative multifunctional structures (Figure 3). A CAD section and top view of the smart tile are shown in Figure 3a,b, respectively. The experimental model top view is shown in Figure 3c.

Figure 2. Aeronautical cellular structures: (**a**) Cellular beam and (**b**) Rectangular cellular panel. Panel thicknesses: top and bottom skin 2 mm, cell width 33 mm, cell height 12 mm, and side walls 1 mm.

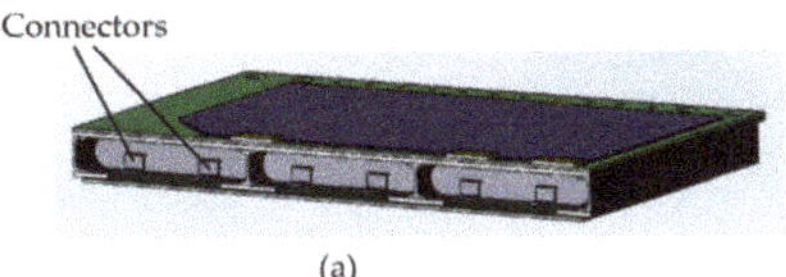

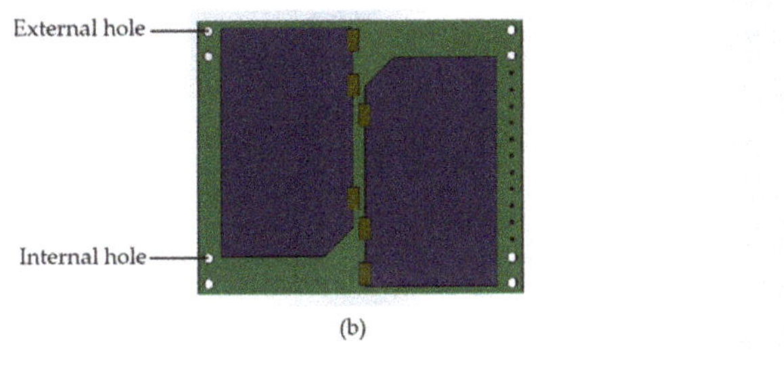

Figure 3. Power management tile: (**a**) CAD model section. The cross-section is colored in white; (**b**) CAD model top view; and (**c**) Experimental model top view.

The cellular configuration has some advantages with respect to the classical configurations in [12] and [16]. It is more rigid of the solid laminate structure [16], thanks to the distance from the neutral axis of the CFRP panels. Moreover, the laminate structure had holes to house the batteries that are not necessary in the cellular structure. The cellular configuration has a major shear and bending stiffness than the sandwich structure [12], thanks to the stiffeners. In addition, CFRP cells obtained with the upper and lower panels and the stiffeners provide protection and containment for the batteries.

2.2.1. Tile Mechanical and Electrical Configuration

The upper and lower structural panels are composite laminates and hold the PCBs for the solar cells (upper panel in Figure 3a) and the electrical subsystem circuitry (lower panel in Figure 3a). The batteries are placed between the two panels, with four CFRP stiffeners among them.

Some electronic details are shown in Figure 4: the magnetorquer (a) is buried in the solar cells PCB. The triple-junction, GaAs solar cells include a protection diode (b) and transmit electrical power through electrical conduits (c). Electrical power is carried to the inner PCB via through-holes (c, d), where it is collected in the battery array.

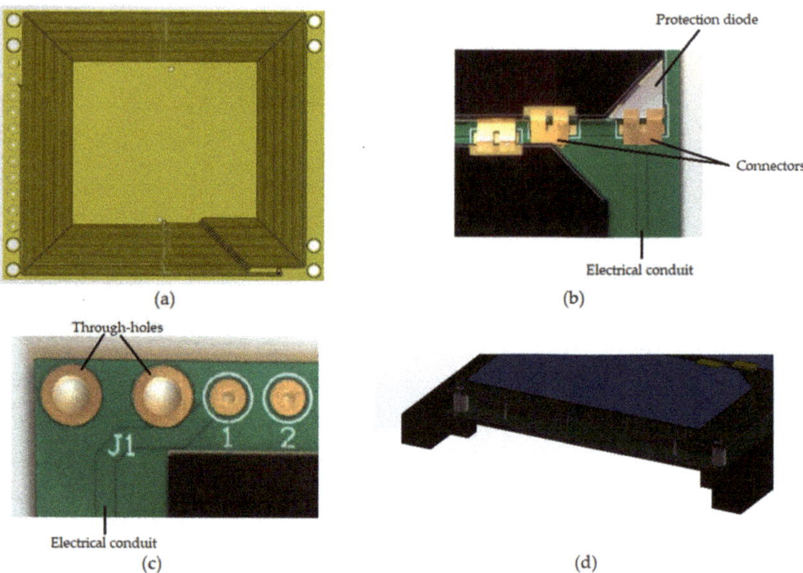

Figure 4. Power management tile electronic details: magnetorquer (**a**), solar cells connectors with protection diode (**b**), solar cells electrical conduit (**c**), and electrical connection with battery array (**d**).

Mechanical connection between the batteries and the CFRP laminates is obtained with a typical silicone resin, to provide thermal protection and vibration damping for the batteries [19]. The resin adheres to all the six sides of the batteries. Electrical connection between the batteries and the inner PCB (lower PCB in Figure 3a) is obtained with holes in the CFRP lower panel to allow the passage of batteries connectors (shown in Figure 3a as thin metal strips). The mechanical connection of the tile with the commercial aluminum structure is obtained with four screws passing either through the external, cylindrical holes or through the internal, cylindrical holes (Figure 3b,c). The screws are typical aerospace inserts with a diameter of 3 millimeters. The use of external or internal holes is dictated by the choice of the cubic aluminum structure face.

With respect to classical configurations [12,16], there is additional area on the PCBs to contain two solar cells, the set of electronic components and their electrical connections to allow the proper operation of the electrical power subsystem. Moreover, it has been proven that lithium batteries, once mounted on CubeSat external panels, offer additional radiation shielding and thermal regulation [20]. However, it has been proven that radiation has negative effects on batteries performance [21].

Protection against space debris has to be studied yet. However, there is a reasonable level of redundancy due to the adoption of six batteries. Although a degradation of the electrical system performance inevitably happens in the case one battery malfunctions, the remaining five can be designed to be electrically independent.

The production of the CFRP structure can be made with standard industrial procedures and is the object of current work. Aluminum molds can be employed with a hand lay-up process of standard prepregs. The assembly then undergoes an autoclave cure cycle. PCBs and batteries are mounted after the polymerization.

2.2.2. Commercial Batteries Selection

Once the Lithium Polymer (LiPo) battery was chosen as the battery type, a preliminary investigation was performed among commercial batteries. Since AraMis is not payload-specific, the objective was to find the maximum energy storage capacity, to accommodate the widest possible range of payloads.

The maximum battery thickness was set to 4.5 mm. This value has been set with the help of the CAD model (Figure 3a,b), considering the total thickness of the stack including one battery, the CFRP laminates, the PCB for the solar cells and its electronic components, once mounted on the commercial structure (Figure 1b). The maximum thickness of the stack is 6.5 mm from the plane of the rails, with an appropriate tolerance, due to CubeSat mechanical requirements discussed in Section 2. Various prismatic batteries have been considered, and the potential tile energy storage capacity was estimated. The selected battery is a Batimex LP452540 with a typical capacity of 1776 mWh (details are given in Table 2). The chosen array of six commercial batteries allows to store approximately 11 Wh in one power management tile. The total capacity is in line with actual commercial 1U-2U EPS systems [22]. It is known that batteries with similar dimensions and superior capacities exist [20,23]. However, these batteries are not COTS and have a considerable cost, thus they do not comply with AraMis project requirements. Since their shape is prismatic, it would be reasonably simple to integrate them in the present design, if they will become suitable for the project in the future. One of the objectives of present work is to demonstrate the feasibility of CubeSat subsystems with COTS components.

Table 2. Selected battery characteristics.

Parameter	Value	
Nominal voltage	3.7	V
Typical capacity	1.78	Wh
Maximum length	40.0	mm
Maximum width	25.5	mm
Maximum thickness	4.5	mm
Typical mass	10.0	g

To increase the EPS overall capacity, the smart tiles modular approach can be exploited to obtain higher energy capacities. With two or three power management tiles mounted on the same 1U CubeSat it is possible to reach 22 or 33 Wh, respectively. Moreover, in multiples of 1U, the area of some smart tiles is increased and thus more commercial batteries can be stored in the same power management tile. For example, a 2U tile has twice the area of a 1U tile, and this allows to store up to 22 Wh with the present design. Moreover, the above-mentioned optimization can be repeated to obtain a major energy capacity per unit of tile surface.

2.3. Structural Analysis

A set of Finite Element (FE) analyses has been performed to assess the compatibility of the design with the launch mechanical environment. The Vega launch vehicle has been chosen for the analysis. The applicable launch condition [24] imposes a Limit Load (LL) of 7 g. The adopted Safety Factor (SF) value is 1.8. It is known that other authors adopted lower safety factors, e.g., Ampatzoglou et al. chose 1.25 [8]. However, with the intention of applying this technology to aeronautics also, a safety factor of 1.5 with an additional special factor of 1.2 has been chosen, to take into account the variability of material properties, as suggested for example by CS-VLA 619 and AMC VLA 619. As a result, the Ultimate Load (UL) becomes:

$$UL = SF \times LL = 1.8 \times 7\,g = 12.6\,g \approx 13\,g. \tag{1}$$

The 13 g acceleration can be applied in any direction, due to the arbitrary position of the tile on the aluminum frame and to the arbitrary orientation of the CubeSat inside the deployer. The worst case condition has been found to happen with acceleration orthogonal to the tile plane and will be described in the results section. The power management tile is mounted on the aluminum alloy frame with screws passing either through the external holes or through the internal holes (Figure 3b). Both conditions have been considered in the analysis. The degrees of freedom of nodes belonging to fastened holes have been preliminarily imposed to be zero. The FE model (Figure 5a) considers the contact of the above mentioned components, i.e., batteries, resin, CFRP laminates and PCBs. Mechanical properties of the batteries cannot easily be estimated, thus values similar to the ones found in the open literature [16] have been adopted. The mechanical properties of the materials can be found in Table 3. The CFRP panels are carbon/epoxy where a carbon fiber fabric is employed, the lamination is [±45]$_s$.

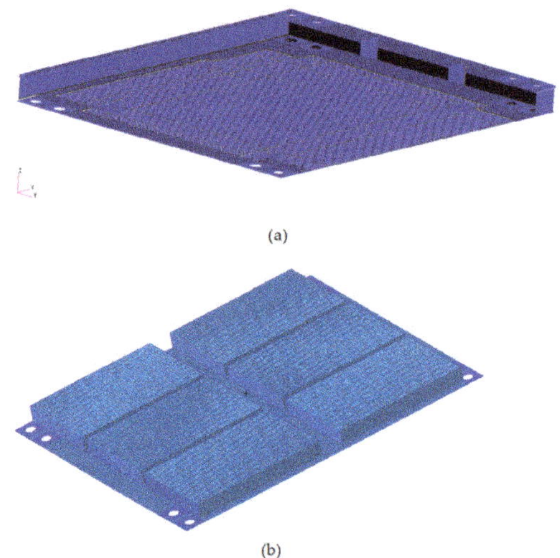

Figure 5. FE models mesh: (**a**) Cellular configuration; (**b**) Commercial Electrical Power System (EPS) architecture.

Table 3. Materials mechanical properties for the Finite Element (FE) analysis.

Material	E [MPa]	ν -	G [MPa]	ϱ [g/cm^3]
Battery assembly	150.00	0.30	57.69	2.08
Resin	218.00	0.30	83.85	1.27
FR4	3000	0.39	1079.14	1.20

Material	E$_{11}$ [MPa]	E$_{22}$ [MPa]	ν$_{12}$ -	G$_{12}$ [MPa]	ϱ [g/cm^3]
CFRP	70000	70000	0.10	5000	1.60

The CFRP panels and stiffeners have been discretized with Quad4 elements with sizes ranging from 0.25 to 0.80 mm. The resin and the batteries have been discretized with Tet4 and Hex8 elements, respectively, with sizes ranging from 0.80 to 1.68 mm.

It has been observed that the mechanical layout of commercial EPSs [22,23] is rather common among commercial solutions and thus a similar configuration has been analyzed. As shown in Figure 5b, the commercial architecture is simpler than the proposed one and consists just of a PCB and six batteries, of the same type. The thickness of the PCB has been increased with respect to the proposed solution, according to commercial EPS architectures.

The cellular and commercial structures have been analyzed with the MSC Nastran™ software [25] and the results have been compared. a static analysis has been conducted to evaluate strains, stresses and maximum displacements. In addition, the natural frequencies and modes have been evaluated.

Simple, preliminary thermal analyses results have shown that the temperature of the batteries is below the maximum operating temperature of 60 °C, a detailed thermal analysis and experimental tests are the object of future work. At least two conditions have to be considered. The first is during the launch, when the launch vehicle ejects the payload fairings. The second is along the nominal Low

Earth Orbit (LEO) orbit, and must consider as inputs the irradiation from the Sun, the terrestrial albedo and internal electrical dissipations.

3. Results and Discussion

3.1. Mass Breakdown

The mass breakdown is shown in Table 4. The mass of the filling can be improved, being almost equal to the mass of the batteries. This aspect needs to be improved and is the object of future studies. The ratio between the total tile mass and the CubeSat maximum mass is 10%. Although it is a partial result and does not involve the whole CubeSat with its payload, it is lower than the structural mass ratio of 30%, identified as unsatisfactory in the STRaND-1 project [6]. There is still margin to place the remaining subsystems components.

Table 4. Mass breakdown for the proposed design.

Component	Mass	
	[g]	[%]
CFRP	15	10.95%
Batteries	60	43.80%
PCBs	7.6	5.55%
Filling	54.4	39.70%
Total	137	100%

A comparison of total and partial masses can be done with a commercial embedded EPS [22] and a commercial aluminum alloy structure [26] (Table 5). For the latter structure, one face of the 1U cube has been considered. The PCBs weight of the proposed design is lower than the commercial embedded EPS (respectively, 7.6 g and 25.6 g). Clearly, in the proposed design there is no intended structural function of the PCBs, and thus their thickness can be reduced, while for the commercial EPS, the PCB panel is the only component able to withstand loads, and thus it has a major thickness. The structural mass for the present design (15 g) is given by the mass of CFRP elements. It is clearly lower than the structural mass of the commercial aluminum alloy panel (28 g), due to the use of composite materials.

Table 5. Masses comparison.

Quantity		Proposed CFRP Design	Commercial Embedded Battery [22]	Commercial Structural Al Panel [26]
Total mass	[g]	137	80	28
PCB mass	[g]	7.6	25.6 (estimated)	-
Structural mass	[g]	15	-	28
Capacity	[Wh]	11	11	0

However, the overall mass of the proposed design is greater than the mass of the commercial EPS. This is mainly due to the resin employed for mechanical and thermal insulation of the batteries, whose mass can be reduced with lighter materials. Nevertheless, in the author's opinion, the present design has advantages with respect to its commercial counterpart. The CFRP cells provide containment and protection from the batteries for the rest of the spacecraft, and future experimental qualification tests are aimed at evaluating the protection of batteries from vibrations, excessive temperatures and thermal shocks. Furthermore with two, lighter PCBs, more area can be devoted to EPS electronic components and circuits. In addition, no internal volume is subtracted to the payload. Finally, the proposed design includes additional components with respect to the commercial design [22], i.e., solar cells, magnetorquers and additional electronic circuitry.

3.2. Finite Element Analysis

The FE static analysis (Figure 6) has revealed that the structure of the smart tile can easily survive the launch, with a maximum displacement of 4.8 μm in the worst case, i.e., with external holes blocked. The results are given in Table 6. As expected, the maximum displacement with the external holes blocked is major of the maximum displacement with internal holes blocked (respectively, 4.8 and 4.6 μm).

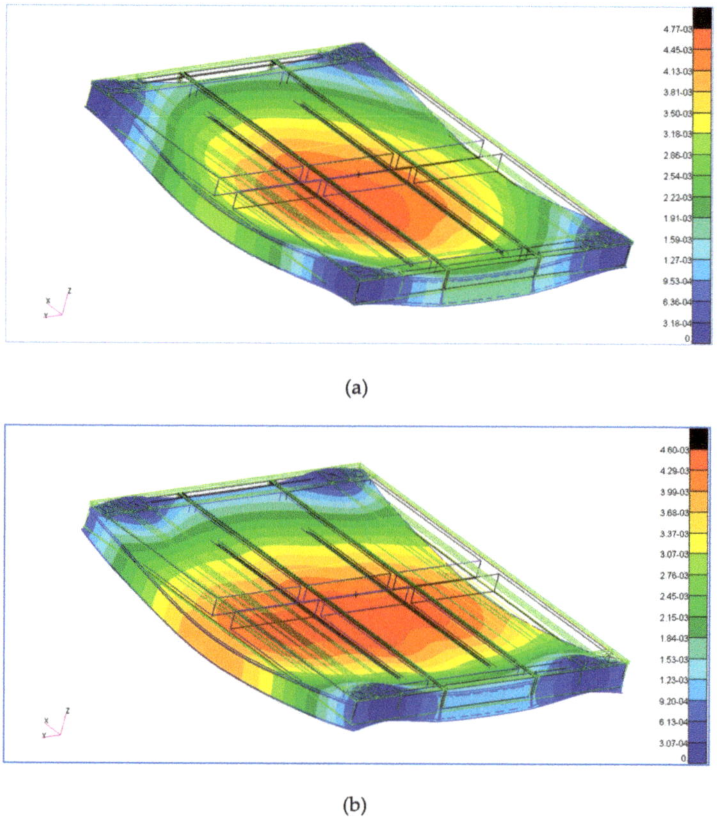

(a)

(b)

Figure 6. Static deformations: (**a**) External holes blocked and (**b**) Internal holes blocked. Displacements are given in millimeters.

Table 6. FE static and modal analysis results.

Quantity		External Holes Blocked	Internal Holes Blocked
Maximum displacement	[μm]	4.8	4.6
Fundamental frequency	[Hz]	963	956

All the stresses and the strains are far below materials allowable limits, with external and internal holes blocked. The maximum stresses on the batteries and PCBs are in the order of tens and hundreds of kilopascals, respectively. These values are far below the materials maximum acceptable stresses.

The modal analysis (Figure 7) has provided fundamental frequencies in the order of 900 Hz for both constraint cases. Vega User's Manual [24] prescribes that the lateral axis fundamental frequency, f_{lat}, must be equal or greater than 15 Hz:

$$f_{lat} \geq 15 \text{ Hz.} \tag{2}$$

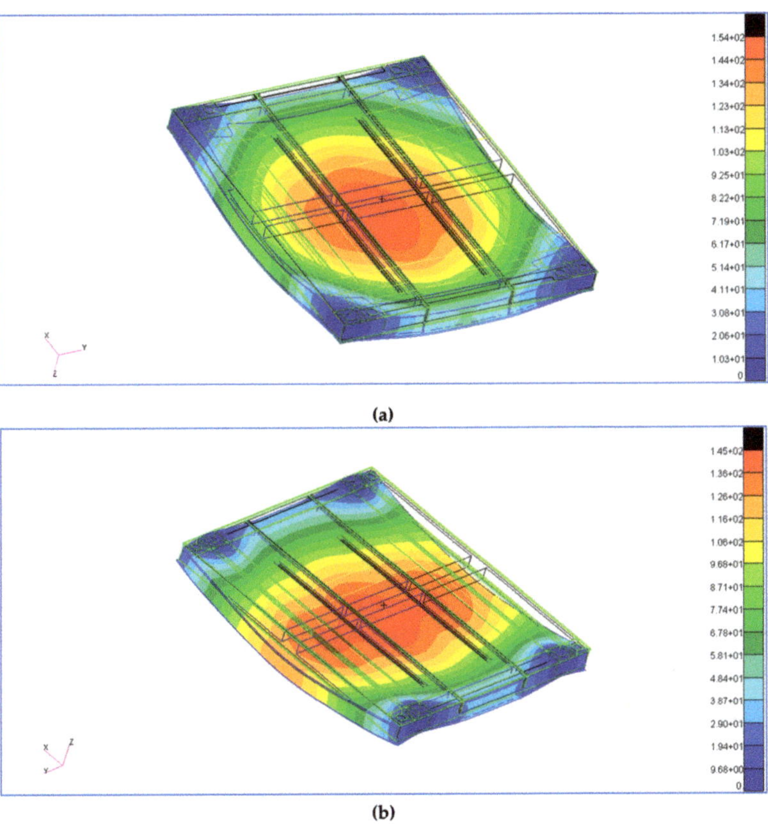

Figure 7. Fundamental modes: (**a**) External holes blocked; and (**b**) Internal holes blocked. Displacements are given in millimetres.

And the longitudinal axis fundamental frequency, f_{long}, must be in the range:

$$20 \text{ Hz} < f_{long} < 45 \text{ Hz or } f_{long} > 60 \text{ Hz.} \tag{3}$$

Although this is a partial result about power management tile local resonance, the above requirements are satisfied. The requirements of Equations 2 and 3 will have to be verified on the whole CubeSat.

At moment the FEM model does not consider details of electrical connections between the PCBs and the batteries. The vibration resistance of electrical connections will be proven during experimental qualification tests. Future work will establish the amount of damage accumulated in the batteries and in their electrical connections due to the launch mechanical environment.

3.3. Comparison with Commercial Architecture

The FE static and modal analysis has been repeated with the mechanical architecture of the commercial EPS (Figure 5b). Results are summarized and compared with the proposed design in Table 7. The advantages of the proposed design are evident, since its maximum displacement is minor and the fundamental frequency is major than the commercial EPS.

Table 7. Comparison between proposed and commercial architectures.

Quantity		Proposed Design	Commercial EPS
		External Holes Blocked	
Maximum displacement	[mm]	4.8×10^{-3}	0.27
Fundamental frequency	[Hz]	963	133

3.4. Comparison with Similar Studies on Embedded Batteries Structures

Besides the structural results comparison with the commercial EPS for CubeSats (Section 3.3), wider comparisons can be made with similar systems for space [20,22,23], automotive [17], and marine [12] environments.

Regarding the CubeSat EPS systems [20,22,23], it is evident that the energy capacity is higher, for some models, than the capacity of the EPS under discussion (11 Wh for the present case, against 22 Wh for [20,23]). However, in addition to the batteries, the proposed design includes solar cells, mounted on the outer PCB, additional electronic circuitry, mounted on the inner PCB, and some versions of the tile include magnetic torquers (Table 1) within the PCBs, in order to provide, together with the other tiles, a spacecraft bus as complete as possible and not only the power storage function. These components were not included in the other battery systems [20,22,23]. Moreover, the present solution is designed to be mounted on the aluminum frame of the CubeSat, to leave internal volume available for the payload, while the other battery systems [20,22,23] are accommodated inside the CubeSat.

Solutions presented in [12,17] have a different structural configuration, i.e., solid beams [12] and sandwich beams [12,17], so a comparison of the structural performances would not be useful. In any case, for both systems the functional integration of the electrical energy storage did not degrade the structural performance below acceptable limits. On the contrary, the bending performances were only marginally affected by the inclusion of the batteries. The present study confirms that, if the system is properly designed, acceptable bending properties can be obtained (Table 6). Although it has been shown that resonance frequencies are affected by the batteries [16], their introduction did not lower the fundamental frequency below acceptable limits (Section 3.2), thanks to the adoption of CFRP.

3.5. Mockup

The geometric compliance of the proposed design has been tested with the production of a mockup of the tile (Figure 8). The CFRP laminates have been 3D printed and the PCB boards have been produced.

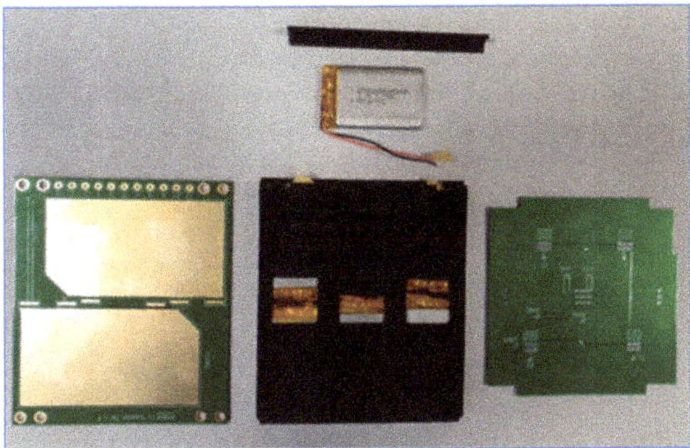

Figure 8. Mockup of the proposed design.

The dimensional congruence of the CFRP structure has been verified by inserting the chosen batteries in their slots and by stacking the CFRP structure and the PCBs in the correct position. A final check was made to assure the compliance of the assembly with the commercial aluminum structure.

4. Conclusions

The results given by the illustrated design method are encouraging. Unlike commercial EPSs [20,22,23], the presented system is mounted on the CubeSat aluminum frame, leaving the internal cubic volume for the payload. The mass of the system is major than equivalent actual commercial EPSs [22], although a way to reduce it has already been foreseen, as described in Section 3.1. Moreover, the presented system includes additional components with respect to commercial EPSs [20,22,23], i.e., magnetic torquers, solar cells and additional circuitry. The FE analysis showed that the structure of the embedded battery can resist to the launch phase maximum acceleration, aboard the Vega launch vehicle. The fundamental resonance frequencies are far from the forbidden intervals. The study of the mockup and the production of the PCBs suggests that the tile is technologically feasible and that it is compliant with the low cost requirement of the AraMiS project.

In addition, the structural analysis hints that this design can be further improved, for example with cut-outs in the carbon fiber laminates. Thus, structural mass would be further reduced, retaining an adequate structural response.

The results seem to indicate that the integration of the spacecraft bus functions in the primary structure can be feasible for nanosatellites. The structure can globally withstand the launch environment, and future work will investigate local effects, e.g., the electrical connections of the batteries. The adoption of composite materials led to a functionally integrated spacecraft bus. Integration of commercial batteries increased the volume available for the payload at a reasonable cost. Volume availability for the payload, as already shown [13], has the priority on mass saving. The availability of mass and volume for the spacecraft payload will be more and more important in the future, with the increasing complexity and the ambitious objectives of incoming space missions.

Author Contributions: Conceptualization, G.C., E.C., L.M.R., and G.R.; formal analysis, G.C. and E.C.; investigation, G.C., E.C., L.M.R., and G.R.; writing—original draft preparation, G.C.; writing—review and editing, G.C., E.C., L.M.R., and G.R.; supervision, G.C., E.C., L.M.R., and G.R. All authors have read and agreed to the published version of the manuscript.

Funding: This research received no external funding.

References

1. Heidt, H.; Puig-Suari, J.; Moore, A.; Nakasuka, S.; Twiggs, R. CubeSat: a new Generation of Picosatellite for Education and Industry Low-Cost Space Experimentation. In Proceedings of the 14th Annual/USU conference on small satellites, Logan, UT, USA, 21–24 August 2000.
2. CubeSat Design Specification. Available online: https://static1.squarespace.com/static/5418c831e4b0fa4ecac1bacd/t/56e9b62337013b6c063a655a/1458157095454/cds_rev13_final2.pdf (accessed on 22 November 2019).
3. Poghosyan, A.; Golkar, A. CubeSat evolution: Analyzing CubeSat capabilities for conducting science missions. *Progr. Aerosp. Sci.* **2017**, *88*, 59–83. [CrossRef]
4. CubeSats Database. Available online: https://sites.google.com/a/slu.edu/swartwout/home/cubesat-database/census (accessed on 22 November 2019).
5. Janson, S.W.; Welle, R.P.; Rose, T.S.; Rowen, D.W.; Hinkley, B.S.; La Lumondiere, S.D.; Maul, G.A.; Werner, N.I. The NASA Optical Communication and Sensors Demonstration Program: Preflight Update. In Proceedings of the 29th Annual AIAA/USU Conference on Small Satellites, Logan, UT, USA, 13 August 2015.
6. Kenyon, S.; Bridges, C.P.; Liddle, D.; Dyer, R.; Parsons, J.; Feltham, D.; Taylor, R.; Mellor, D.; Schofield, A.; Linehan, R. STRaND-1: Use of a $500 Smartphone as the Central Avionics of a Nanosatellite. In Proceedings of the 62nd international astronautical congress, Cape Town, South Africa, 3–7 October 2011.
7. Pranajaya, F.M.; Zee, R.E. Generic Nanosatellite Bus for Responsive Mission. In Proceedings of the 5th responsive space conference, Los Angeles, CA, USA, 23–26 April 2007; AIAA-RS5 2007-5005.
8. Ampatzoglou, A.; Baltopoulos, A.; Kotzakolios, A.; Kostopoulos, V. Qualification of Composite Structure for Cubesat Picosatellites as a Demonstration for Small Satellite Elements. *IJASAR* **2014**, *1*, 1–10.
9. Speretta, S.; Reyneri, L.M.; Sanso´e, C.; Tranchero, M.; Passerone, C.; Corso, D.D. Modular Architecture for Satellites. In Proceedings of the 58th International Astronautical Congress, Hyderabad, India, 24–28 September 2007.
10. Johannisson, W.; Ihrner, N.; Zenkert, D.; Johansson, M.; Carlstedt, D.; Asp, L.E.; Sieland, F. Multifunctional performance of a carbon fiber UD lamina electrode for structural batteries. *Compos. Sci. Technol.* **2018**, *168*, 81–87. [CrossRef]
11. Adam, T.J.; Liao, G.; Petersen, J.; Geier, S.; Finke, B.; Wierach, P.; Kwade, A.; Wiedemann, M. Multifunctional Composites for Future Energy Storage in Aerospace Structures. *Energies* **2018**, *11*, 335. [CrossRef]
12. Thomas, J.P.; Qidway, M.A.; Pogue III, W.R.; Rohatgi, A. Multifunctional structure-battery composites for marine systems. *J. Compos. Mater.* **2013**, *47*, 5–26. [CrossRef]
13. Swartwout, M. The First One Hundred CubeSats: a Statistical Look. *JoSS* **2013**, *2*, 213–233.
14. Romeo, G.; Borello, F.; Correa, G.; Cestino, E. ENFICA-FC: Design of transport aircraft powered by fuel cell & flight test of zero emission 2-seater aircraft powered by fuel cells fueled by hydrogen. *Int. J. Hydrogen Energ.* **2013**, *38*, 469–479.
15. Romeo, G.; Frulla, G.; Cestino, E. Design of a High-Altitude Long-Endurance Solar-Powered Unmanned Air Vehicle for Multi-Payload and Operations. *Proc. Inst. Mech. Eng. G* **2007**, *221*, 199–216. [CrossRef]
16. Galos, J.; Afaghi Khatibi, A.; Mouritz, A.P. Vibration and acoustic properties of composites with embedded lithium-ion polymer batteries. *Compos. Struct.* **2019**, *220*, 677–686. [CrossRef]
17. Galos, J.; Best, A.S.; Mouritz, A.P. Multifunctional sandwich composites containing embedded lithium-ion polymer batteries under bending loads. *Mater. Design* **2020**, *185*. [CrossRef]
18. Romeo, G. Sailplane wing box design by use of graphite/aramide/epoxy material. *Tech. Soar.* **1981**, *2*, 70–75.
19. The Use of Silicone Adhesives in Space Applications. Available online: https://www.adhesivesmag.com/articles/85082-the-use-of-silicone-adhesives-in-space-applications (accessed on 25 November 2019).
20. Nader, R.; Uriguen, M.; Drouet, S.; Nader Drouet, G. High Energy Density Battery Array for CubeSat Missions. In Proceedings of the 67th International Astronautical Congress, Guadalajara, Mexico, 26–30 September 2016; Curran Associates, Inc.: New York, NY, USA, 2017; p. 32401.
21. Tan, C.; Lyons, D.J.; Pan, K.; Yee Leung, K.; Chiurazzi, W.C.; Canova, M.; Co, A.C.; Cao, L.R. Radiation effects on the electrode and electrolyte of a lithium-ion battery. *J. Power Sources* **2016**, *318*. [CrossRef]
22. Crystalspace P1U "Vasik" EPS. Available online: https://www.cubesatshop.com/product/crystalspace-p1u-vasik/ (accessed on 22 November 2019).

23. BA0x High Energy Density Battery Array. Available online: https://www.cubesatshop.com/product/ba0x-high-energy-density-battery-array/ (accessed on 26 November 2019).
24. Vega User's Manual Issue 4 Revision 0. Available online: http://www.arianespace.com/wp-content/uploads/2015/09/Vega-Users-Manual_Issue-04_April-2014.pdf (accessed on 22 November 2019).
25. MSC Nastran™ Quick Reference Guide 2017. Available online: https://simcompanion.mscsoftware.com/infocenter/index?page=content&id=DOC11146& (accessed on 11 February 2020).
26. CubeSat KitTM Pro Chassis Walls. Available online: https://www.pumpkinspace.com/store/p42/CubeSat_Kit%E2%84%A2_Pro_Chassis_Walls.html (accessed on 22 November 2019).

Article

Estimation of Performance Parameters of Turbine Engine Components Using Experimental Data in Parametric Uncertainty Conditions

Olexandr Khustochka [1], Sergiy Yepifanov [2], Roman Zelenskyi [2] and Radoslaw Przysowa [3,4,*]

[1] SE Ivchenko-Progress, 69068 Zaporizhia, Ukraine; a.khustochka@ivchenko-progress.com
[2] Aircraft Engine Department, National Aerospace University, "Kharkiv Aviation Institute", 61070 Kharkiv, Ukraine; s.yepifanov@khai.edu (S.Y.); r.zelenskyi.roman@gmail.com (R.Z.)
[3] Instytut Techniczny Wojsk Lotniczych (ITWL), ul. Księcia Bolesława 6, 01-494 Warszawa, Poland
[4] Technology Partners, ul. Pawińskiego 5A, 02-106 Warszawa, Poland
* Correspondence: radoslaw.przysowa@itwl.pl

Received: 30 November 2019; Accepted: 10 January 2020; Published: 16 January 2020

Abstract: Zero-dimensional models based on the description of the thermo-gas-dynamic process are widely used in the design of engines and their control and diagnostic systems. The models are subjected to an identification procedure to bring their outputs as close as possible to experimental data and assess engine health. This paper aims to improve the stability of engine model identification when the number of measured parameters is small, and their measurement error is not negligible. The proposed method for the estimation of engine components' parameters, based on multi-criteria identification, provides stable estimations and their confidence intervals within known measurement errors. A priori information about the engine, its parameters and performance is used directly in the regularized identification procedure. The mathematical basis for this approach is the fuzzy sets theory. Synthesis of objective functions and subsequent scalar convolutions of these functions are used to estimate gas-path components' parameters. A comparison with traditional methods showed that the main advantage of the proposed approach is the high stability of estimation in the parametric uncertainty conditions. Regularization reduces scattering, excludes incorrect solutions that do not correspond to a priori assumptions and also helps to implement the gas path analysis with a limited number of measured parameters. The method can be used for matching thermodynamic models to experimental data, gas path analysis and adapting dynamic models to the needs of the engine control system.

Keywords: gas turbine engine; performance model; gas path analysis; robust estimation; identification; regularization; fuzzy set; membership function

---

1. Introduction

Aircraft gas turbine engines are the most expensive and vital components of the aircraft. Their maintenance and operating costs depend on the engine performance, which is mainly determined by a condition of the engine gas path components such as the compressor, turbine, combustor, ducts, etc. For this reason, engine maintenance is usually supported by a prognostics and health management system (PHM), which usually implements a diagnostic procedure known as gas path analysis (GPA).

Nowadays, different approaches are used for processing flight and test-cell data to estimate component performance and detect faults [1]. These methods can be divided into two groups [2]:

- Model-based methods which use thermodynamic models to relate failures to measured parameters of the gas path [3]
- Artificial intelligence (AI) methods which belong to expert systems [4]

Model-based and hybrid systems [5] are effective only if a validated engine model is available.

The non-linear thermo-gas-dynamic models of turbine engines [6] and identification procedures have been applied in diagnostics for more than 40 years. Identification adjusts the model to make its output parameters as close as possible to the experimental data [7–14]. Besides the significant improvement in the gas path simulation, the estimated parameters contain information about the health of each component. Model identification involves the minimization of a functional which is a sum of squared deviations, calculated as differences of calculated and measured values. This is the Least Squares Method (LSM) which is usually implemented in non-linear GPA solvers using iterative Newton–Raphson algorithm.

Matching the engine model to experimental data is characterized by the presence of multiple parameters, which can be used for the model correction. These parameters can be strongly correlated. At the same time, a quantity of measured parameters is strongly limited. These reasons decrease the stability of the correction procedure, which is based on LSM, and forced researchers to look for methods to improve the stability. Hence, matching engine models needs regularization.

For more stable solutions, V. Borovick and Ye. Taran applied the least modulus method [15], which is more robust to outliers. S. Yepifanov implemented the Levenberg–Marquardt method [16] as well as Singular Value Decomposition and ε-structuration [17,18]. A. Volponi et al. applied the Kalman filter [19], and this approach is followed and developed in many papers [20–24], whose authors improved stability of the algorithm and its applicability to a non-linear engine model. X. Chang et al. applied an alternative method based on the non-linear filtration (sliding mode observer) [25–27].

Stability of identification procedures can be improved by regularization. This tool has been long known and used in statistical parameter estimation but was not common in system identification until recently [28]. T. Breikin et al. presented a regularization-based approach to the estimation of engine model coefficients [29]. It was based on the regularity of the frequency response pattern over the operation range of the engine, but that method is not universal. This paper is based on the Levenberg–Marquardt algorithm and Tikhonov regularization which were also used by S. Guseynov et al. [30,31]. These papers did not analyse the bias nor consider how the regularization coefficient influences the bias and the noise, which could impede practical applications.

It is well-known that regularization reduces the modification of the LSM functional by adding the regularizing component to it. This causes biased estimates of the model coefficients. B. Roth, D. Doel et al. demonstrated this clearly [32]. The biased estimates cause errors in GPA diagnosing solutions. The errors of the estimated parameters caused by the bias need to be monitored, but the traditional methods of regularization essentially prevent this. There is no universal recommendation on a choice of the weight coefficient for the added regularizing component.

This paper presents a new method for the regularized identification of gas-turbine models based on a priori information (Figure 1). First, the traditional Tikhonov regularisation is applied to simulated engine data to analyse the impact of regularization on a non-linear least squares solution and select the optimal value of the regularisation coefficient (Section 2.3). Next, membership function is used to implement a priori information into non-linear identification procedure, which is based on a generic algorithm (Section 2.4). Finally, the proposed method is tested with a turboshaft model and test-cell data and compared with traditional approaches.

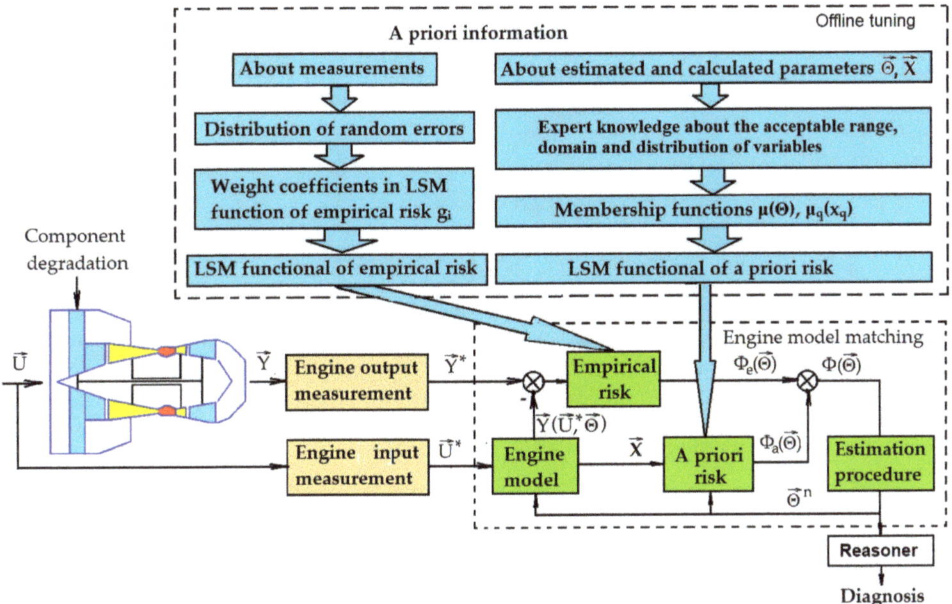

Figure 1. Regularized multi-criteria identification of gas-turbine model.

The proposed approach is a significant modification of the standard model-based procedure for estimation of the performance parameters of engine components. The ability to implement prior knowledge in different forms to stabilize the solution is the main contribution of the paper. Regularized multi-criteria identification enables engine users to implement Non-Linear GPA [1,2] in ill-conditioned configurations with a limited number of measured parameters.

2. Materials and Methods

2.1. Basic Identification Procedure

The model calculates parameters of the engine gas path $\vec{Y}$ at steady-state modes depending on an operating point, external conditions $\vec{U}$ and component performance $\vec{\Theta}$. Hence, in the general case, it is presented as:

$$\vec{Y} = F(\vec{U}, \vec{\Theta}) \tag{1}$$

The linear model may be formulated as

$$\delta\vec{Y} = H\delta\vec{\Theta} \tag{2}$$

which relates small deviations of the gas path parameters $\delta\vec{Y}$ and parameters of components' performances $\delta\vec{\Theta}$ at a single operating point. H is an influence coefficient matrix (ICM). The linear model (2) is a component of the identification algorithm for the non-linear model.

Identification of the non-linear model (1) by measuring results $\vec{Y}^*$ and $\vec{U}^*$ requires determining estimations $\hat{\vec{\Theta}}$, which are solutions of the following optimization task:

$$\Phi(\vec{\Theta}) = \|\vec{Y}^* - \vec{Y}(\vec{U}^*, \vec{\Theta})\|, \quad \Phi(\hat{\vec{\Theta}}) = \min\Phi(\vec{\Theta}) \tag{3}$$

where $\|\|$ is the Euclidean norm of the vector, Φ–least squares functional to be minimized.

The precision of estimation will improve if the amount of a priori information is higher than the number of unknown parameters. Therefore, the identification can be provided by a set of measurements that are done at N different operating points. For this purpose, the generalized vector of residuals is minimized:

$$\vec{Z}(\vec{\Theta}) = \begin{bmatrix} \vec{Y}_1^* - Y(\vec{U}_1^*, \vec{\Theta}) \\ \vec{Y}_2^* - Y(\vec{U}_2^*, \vec{\Theta}) \\ \ldots\ldots\ldots \\ \vec{Y}_N^* - Y(\vec{U}_N^*, \vec{\Theta}) \end{bmatrix} \tag{4}$$

Model (1) which is included in (4) is numerical. Therefore, the minimization of residuals (4) is a numerical iterative procedure, which in each iteration determines the solution as a sum of the previous solution and current correction:

$$\vec{\Theta}^{n+1} = \vec{\Theta}^n + \Delta\vec{\Theta}^{n+1} \tag{5}$$

and iterations continue until corrections become negligible.

The correction $\Delta\vec{\Theta}^{n+1}$ is determined as a solution of the overdetermined system of linear algebraic equations:

$$C(\vec{\Theta}^n)\Delta\vec{\Theta}^{n+1} = Z(\vec{\Theta}^n) \tag{6}$$

where

$$C(\vec{\Theta}^n) = \begin{bmatrix} H_1(\vec{\Theta}^n) \\ H_2(\vec{\Theta}^n) \\ \ldots\ldots \\ H_N(\vec{\Theta}^n) \end{bmatrix} \tag{7}$$

is a generalized matrix of influence coefficients, which is composed of elementary matrixes $H_i(\vec{\Theta}^n)$ determined for each operating point. In accordance with Equation (6), for each iteration, such correction $\Delta\vec{\Theta}^{n+1}$ of required parameters is found, which minimizes residuals $\vec{Z}(\vec{\Theta})$ (4).

The known solution of the linear system (6) by the least squares method takes the form:

$$\Delta\vec{\Theta}^{n+1} = A^{-1}C^T G\vec{Z} \tag{8}$$

where $A = C^T GC$–Fisher information matrix, G–weight diagonal matrix, whose elements are inverse to variations of measuring errors $\sigma^2_{y\,i}$.

Unfortunately, the least squares method is sensitive to outliers in the right part of the system (6), which can be related to faults in experimental data. Therefore, practical applications of the classic LSM suffer from the instability of estimations $\Delta\vec{\Theta}^{n+1}$ and poor convergence of the identification algorithm as a whole. In some cases, the estimations $\hat{\Theta}$ are far from the expected values. The reasons for such results may be the lack of empirical information, an excessive number of estimated parameters or correlation between two or more state parameters. This causes ill-conditioning of the Fisher Matrix and excessive deviations of estimations. These estimations lose a physical sense (are out of range of possible components' performances parameters variation—for example, efficiency is more than one) and can cause the model calculation program to crash.

Hence, the LSM identification procedure needs modification to provide its stability and physically adequate estimations.

2.2. Regularized Identification Procedure

In the known regularization procedure named for Tikhonov, the generalized functional is minimized, which besides the residuals of measured parameters includes a norm of the finding parameter vector with a weighting factor α. The identification task takes the following form:

$$\Phi'(\vec{\Theta}) = \sum_{i=1}^{m} g_i \sum_{j=1}^{N} \left[y_i(\vec{U}_j, \vec{\Theta}) - y^*_{ij} \right]^2 + \alpha \sum_{k=1}^{r} \theta^2_k; \Phi'(\hat{\Theta}) = \min \Phi'(\vec{\Theta}) \tag{9}$$

where g_i is the weight coefficient that describes measurement error ($g_i = \frac{1}{\sigma^2_i}$), y_i is a component of vector $\vec{Y}$ and θ_k is a component of vector $\vec{\Theta}$.

The regularized identification is based on the least squares method described above. As it is shown in Equation (9), the minimized functional is extended by elements which are related to parameters $\vec{\Theta}$ to be found. Thus, the main changes in the identification procedure are the generalized vector of residuals $\vec{Z}(\vec{\Theta})$ and generalized matrix of influence coefficients $C(\vec{\Theta})$. The modified terms have the following structure:

$$\vec{Z}'(\vec{\Theta}) = \begin{bmatrix} \vec{Z}(\vec{\Theta}) \\ \alpha\vec{\Theta} \end{bmatrix}, C'(\vec{\Theta}) = \begin{bmatrix} C(\vec{\Theta}) \\ \alpha I \end{bmatrix} \tag{10}$$

where I is an identity matrix.

This identification procedure will deliver the non-regularized solution if the regularization coefficient is zero. Otherwise, the influence of coefficient α depends on the proportion between components: $\sum_{i=1}^{m} g_i \sum_{j=1}^{N} \left[y_i(\vec{U}_j, \vec{\Theta}) - y^*_{ij} \right]^2$ and $\sum_{k=1}^{r} \theta^2_k$ of the generalized functional $\Phi'(\Theta)$. Therefore, the effect of regularization is determined by the conditions of identification such as the number of measured parameters *m*, number of operating points *N*, measuring errors G and number of estimated parameters *r*.

2.3. Numerical Simulation

To assess the impact of the regularization coefficient α, the identification procedure was tested with the zero-dimensional thermodynamic model of a turboshaft engine [17] (Figure 2), based on experimental component maps. Performed numerical experiments involved calculation of N = 12 operating points for two scenarios:

(1) Without measuring noise, for a wide range of α values;
(2) With measuring noise, for selected α values.

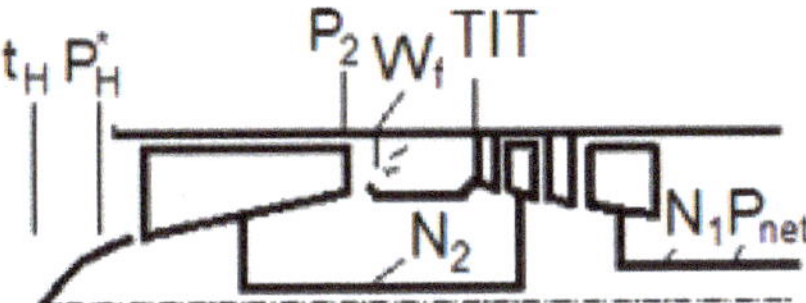

Figure 2. Scheme and measured parameters of the engine.

Five gas path parameters Y_i were used for model identification:

(1) Compressor discharge pressure p_2;
(2) Turbine inlet temperature TIT;
(3) Power turbine discharge temperature;

(4)   Rotational speeds of rotors N_2 and N_1.

Initial deviations of two compressor performance parameters were assumed as follows:

- Deviation of mass flow rate: $\Theta_{1\,sim} = \Delta W_C = -0.03$;
- Deviation of compressor efficiency: $\Theta_{2\,sim} = \Delta \eta_C = -0.04$.

Four estimated parameters included:

- Deviation of flow rate: $\Theta_1 = \Delta W_C$;
- Deviation of compressor efficiency: $\Theta_2 = \Delta \eta_C$;
- Deviation of HPT mass flow: $\Theta_3 = \Delta W_{HPT}$;
- Deviation of HPT efficiency: $\Theta_4 = \Delta \eta_{HPT}$.

Figure 3 presents estimations $\hat{\vec{\Theta}}$ and simulation results $\hat{\vec{Y}} = F(\vec{U}, \hat{\vec{\Theta}})$ for different values of α, which varied in a range from 0 to 1. ΔY_{av} is the average deviation between initial and estimated models:

$$\Delta Y_{av} = \sqrt{\frac{1}{mN}\sum_{j=1}^{N}\sum_{i=1}^{m}\left[y_i(\vec{U}_j, \hat{\vec{\Theta}}) - y_i(\vec{U}_j, \vec{\Theta} = 0)\right]^2} \qquad (11)$$

and ΔY_{av}^* is the average deviation between measurements and estimated model:

$$\Delta Y_{av}^* = \sqrt{\frac{1}{mN}\sum_{j=1}^{N}\sum_{i=1}^{m}\left[y_i(\vec{U}_j, \hat{\vec{\Theta}}) - y_{ij}^*\right]^2} \qquad (12)$$

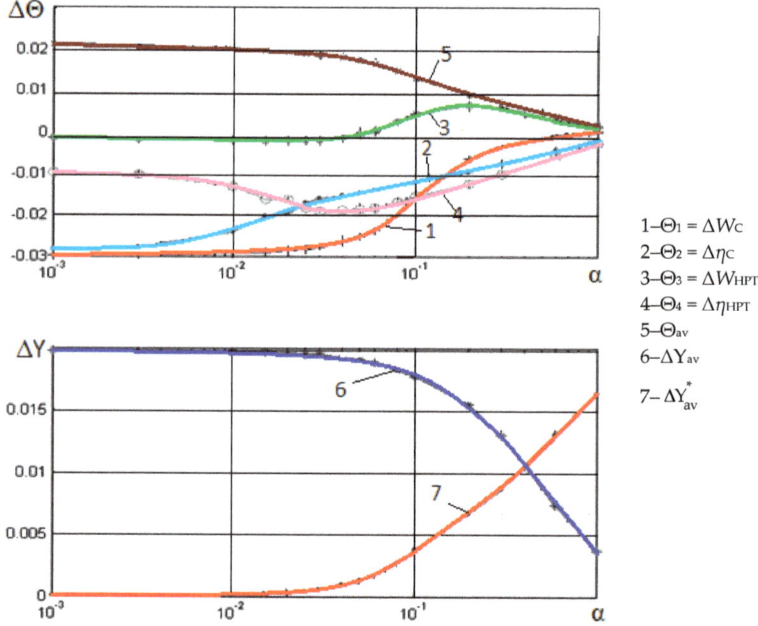

Figure 3. Influence of the regularization coefficient on estimations when noise is absent.

Figure 3 shows that despite the initial growth of Θ_3 and Θ_4 absolute values, the average value Θ_{av} decreases monotonically, whereas the residual between measurements and the corrected model increases. This corresponds to theoretical predictions of the influence of regularization on the identification process.

The simulation results help to choose the value of the regularization coefficient α. If the deviation of parameters $\vec{Y}$ is to be lower than 5 % and the deviation of parameters Θ lower than 1%, then the value of α cannot exceed 0.03.

To obtain precise values of variance and to analyse the distributions of estimations $\hat{\vec{\Theta}}$, a sequence of data sets with random measuring errors with variance $\sigma_{Y_i}^2$ was generated and identified 1000 times. This provided the average precision of about 1% of initial residuals for parameters Y and about 0.5 % of simulated deviation 0.03 for the estimated parameters $\hat{\vec{\Theta}}$.

Figure 4 presents histograms for calculations with two estimated parameters at $\alpha = 0$ and $\alpha = 0.04$. Both non-regularized and regularized estimations have distributions close to normal ones. Total scatter of estimations is also similar, but centres of regularized estimations moved from their true values by 0.008 and 0.005, respectively.

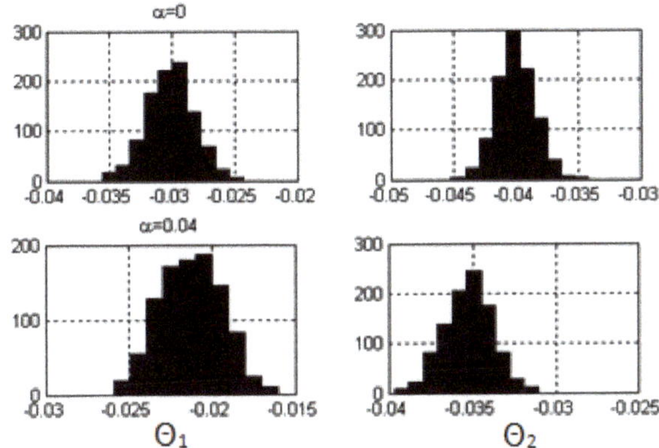

Figure 4. Distributions of estimations when two parameters are estimated.

When four parameters were estimated (Figure 5), estimations $\hat{\Theta}_1$ and $\hat{\Theta}_3$ preserved their distributions while estimations $\hat{\Theta}_2$ and $\hat{\Theta}_4$ had significant differences. Such abnormal behaviour of parameters Θ_2 and Θ_4 is explained by their correlation.

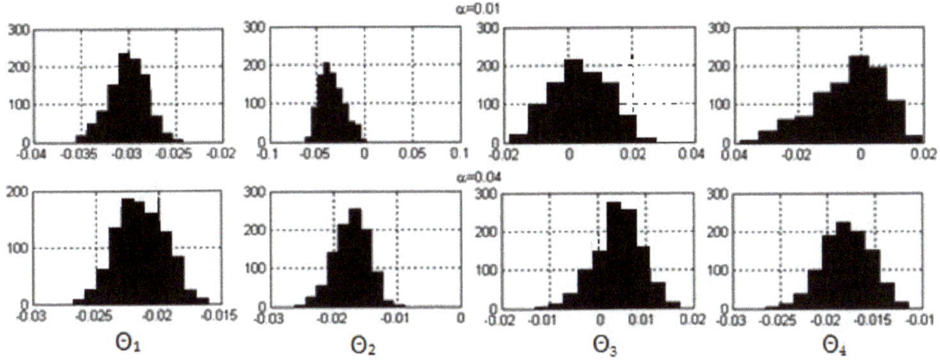

Figure 5. Distributions of estimations when four parameters are estimated.

At low values of the regularization coefficient α, centres of distributions are close to their true values, but scatter is high, and distributions are strongly asymmetrical. When α increases, then mean values move, but scatter decreases, and its distribution approaches the normal one.

As shown above, the value of the regularization coefficient α has to be set appropriately to provide meaningful results of regularized least squares. This procedure is useful in ill-conditioned engine model fitting but needs preliminary adjustment and sensitivity analysis.

2.4. Regularized Identification Procedure Development Using a Priori Information

This paper demonstrates an idea to implement engine heuristics involving component parameters and performances to improve the precision and stability of model identification. The main difficulty is in the diversity of this information, which is presented in one of the following forms:

- Exact statement (for example, a part-load performance in a determined area is smooth);
- Statement in the form of limitations of the area of acceptable solutions (for example, the efficiency of the individual compressor cannot differ more than 3% from the efficiency of an "average" compressor, the performance of which is used in the initial model);
- Statement in the form of fuzzy information (for example, the gas temperature in the turbine will grow with the engine's life);
- Statistical form (for example, probability density functions of parameters).

The next difficulty is in the formalization of parameters that characterize the model quality. These parameters are set on the basis of subjective preferences of decision makers (DM). The same difficulties appear in the ranking of partial criteria and limitations according to their significance for estimation of the model quality. The analysis showed that the main problem is that theoretical-probabilistic methods are difficult to apply in the presence of uncertainties, which are related to subjective preferences whose nature is not statistical. Actually, the choice of the model's structure is the DM procedure, which in multi-criteria case is inevitably highly subjective. If the complexity of the task increases, the role of quality factors will grow. Therefore, it is possible to take into account all criteria using the proper mathematical tool.

We propose to use the fuzzy set theory [33] as this tool. This theory makes the uniform base to describe the information given in all the forms listed above, thus providing the correct mathematical definition of the identification task.

An example of applying fuzzy sets in the GPA and engine model matching is given by M. Zwingenberg et al. [34]. They used fuzzy logic for the evaluation of sensor failures. In contrast, we introduce a priori information about engine performance and experimental data directly into the stabilizing functional through the fuzzy logic approach.

Generalization of the functional (9) gives:

$$\Phi(\vec{\Theta}) = \Phi_e(\vec{\Theta}) + \Phi_a(\vec{\Theta}) \tag{13}$$

where $\Phi_e(\vec{\Theta})$ is the least squares functional (3), which minimizes measurement error, so it is called the empirical risk functional and $\Phi_a(\vec{\Theta})$ is the stabilizing functional, which is considered as the functional of a priori risk.

The determined a priori information is set as a limitation, which may take the form of an equality or inequality. The more general form of setting a priori information is its representation as a fuzzy set. The fuzzy set of parameter x is represented as a definition domain and a membership function $\mu(x)$ in this domain. For the considered task, the parameters x may be the model parameters $\vec{\Theta}$ or the output (calculated) variables $\vec{X}$.

Next, we will use limited normal fuzzy sets with limited definition domain (x_{min}, x_{max}) and $0 \leq \mu(x) \leq 1$. We will express each a priori information as a particular functional of a priori risk

$\Phi_{a\,q}(\vec{\theta})$ and will determine the general functional of a priori risk as a linear composition of particular functionals:

$$\Phi_a(\vec{\theta}) = \sum_{q=1}^{Q} \alpha_q \Phi_{a\,q}(\vec{\theta}) \tag{14}$$

where α_q are weight coefficients.

Let us consider some types of a priori information and its expression in view of fuzzy sets:

Case 1. Limitations of some parameters are known. For example, it is known that $0 < \eta < 1$ and $0 < \sigma < 1$, etc. Using experience and calculation results, these limits can be significantly reduced; for example, $0.5 < \eta < 0.9$ and $0.9 < \sigma < 0.99$ (Figure 6).

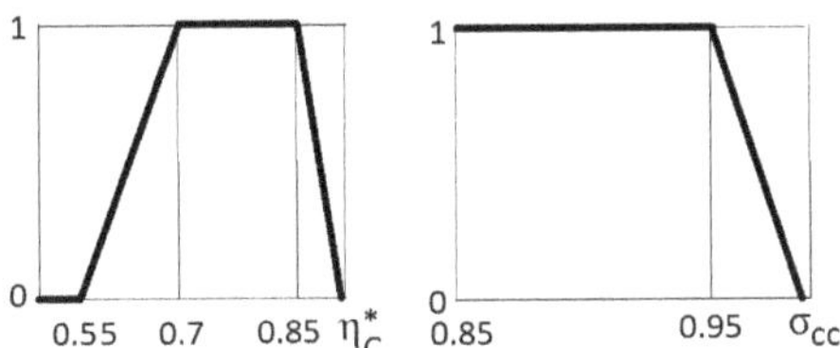

Figure 6. Examples of membership functions.

Case 2. The a priori mathematical model is known. This can be the model with design maps of components or the model of the average engine, which is matched with previous testing results.

This information may be expressed as a relationship between μ_a or Φ_a and the difference between parameters that correspond to the matched and a priori models. These parameters may be the model parameters $\vec{\Theta}$ as measured (for example fuel flow) or non-measured calculated parameters (for example thrust). The membership function, in this case, can be of symmetrical triangular shape and the functional of a priori risk $\Phi_{a\,q}(\vec{\theta})$ that characterizes the similarity between values of the parameter x_q of matched and a priori models can be formed as

$$\Phi_{a\,q}(\vec{\Theta}) = \sum_{j=1}^{N} \mu_q(\Delta x_{q\,j}) \cdot (\Delta x_{q\,j})^2 \quad \Delta x_{q\,j} = x_q(\vec{U}_j,\, \hat{\Theta}) - x_{q\,0}(\vec{U}_j,\, \hat{\Theta}) \tag{15}$$

where x_q is a calculated parameter of the engine; $x_q(\vec{U}_j,\, \hat{\Theta})$ is the value calculated by the model to be matched; $x_{q\,0}(\vec{U}_j,\, \vec{\Theta}_0)$—the value calculated by the a priori model; $\vec{\Theta}_0$ are the parameters of the a priori model.

Case 3. Information about the confidence in different sets of experimental data obtained in different conditions with different precision.

Case 4. Confidence in the available maps of the engine components. For example, we know that the compressor map used in the model corresponds to the old version of the engine and is far from the actual map. Thus, we can express this knowledge in view of the confidence functions that are in this case the membership functions.

The empirical risk functional $\Phi_e(\vec{\Theta})$ in Equation (3) may be considered as a square of Euclidian distance between two sets, one of which contains experimental data, and the other is composed of simulation results:

$$\Phi_e(\vec{\Theta}) = \frac{1}{mN} \sum_{i=1}^{m} g_i \sum_{j=1}^{N} \left[y_i(\vec{U}_j,\, \vec{\Theta}) - y_{ij}^* \right]^2 \tag{16}$$

By analogy, the stabilizing functional can be formed as a second power of a distance from estimated parameter θ_k (or function $x(\vec{\Theta})$ that is determined using the estimated parameters).

The main problem in this analogy is that the fuzzy set that contains a priori information is infinite, so the sum in the above equation must be replaced with integral. For example, the second power of the distance from the engine map parameter $\theta = x$ to the fuzzy set with a given membership function $\mu(\theta)$ equals:

$$\Phi_a(\vec{\theta}) = \frac{\int\limits_{-\infty}^{\infty} \mu(\theta)(x-\theta)^2 d\theta}{\int\limits_{-\infty}^{\infty} \mu(\theta) d\theta} \tag{17}$$

In the iteration process of the task solution, each integral must be calculated numerically. This requires a lot of time and high computational capacity. A numerical solution can be found for certain types of the membership function. We considered the trapezoidal function, the particular cases of which are the rectangle, triangle and isosceles trapezium (Figure 7).

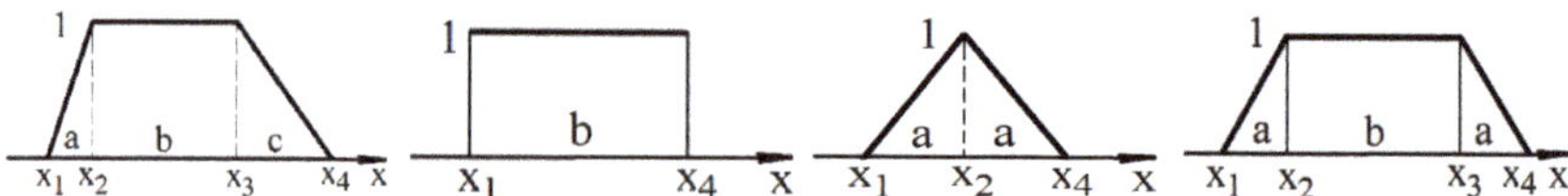

Figure 7. Considered membership functions.

The functional derived for the trapezium:

$$\Phi_{a\,q}(\vec{\theta}) = X_q^2 - \{X_q\big[x_1(a + 2b + c) + \tfrac{1}{3}(2a^2 + 3b^2 + c^2 + 6ab + 3ac + 3bc)\big] + \tfrac{ax_1^2}{2} + \tfrac{2}{3}a^2 x_1 + \tfrac{a^3}{4} + b\big(x_1^2 + 2ax_1 + bx_1 + a^2 + ab + \tfrac{b^2}{3}\big) + \tfrac{c}{12}\big[6x_1^2 + (12a + 12b + 4c)x_1 + 6a^2 + 6b^2 + 12ab + 4ac + 4bc + c^2\big]\}/(\tfrac{a}{2} + b + \tfrac{c}{2}). \tag{18}$$

The proposed modifications of the objective functional cause the non-linearity of the estimation problem, so traditional solution methods are not applicable. This problem is overcome by adapting genetic optimization algorithm to the specifics of the engine model matching [35]. Genetic algorithms are increasingly used in gas turbines to solve complex optimization problems [4,10,36,37].

3. Results

The designed methods were implemented using experimental data obtained during the test-cell testing of the helicopter turboshaft engine. Table 1 contains the values of measured parameters. The last row presents mean squared measurement errors σ_i, which were used to form the diagonal weight matrix G (Equation (8)).

Table 1. Parameters acquired during bench testing.

Operating Mode	t_H, °C	P^*_H, Pa	N_2, %	N_1, %	P_{net}, kW	W_f, kg/h	TIT, K	CPR
1st cruise	12.0	100,192	94.96	94.58	1223	372.5	1017	8.05
2nd cruise	11.6	100,192	97.05	96.87	1521	439.0	1073	8.91
Nominal	12.3	100,192	98.67	98.51	1729	482.1	1106.5	9.41
Maximum	11.9	100,192	102.34	102.02	2203	596.4	1197	10.54
σ_i, %	0.4	0.3	0.085	0.085	0.4	0.3	0.4	0.3

The model parameters to be corrected:

$\pi_C^*(c)$—Scaling factor of CPR (Compressor Pressure Ratio);

$\pi_C^*(a)$—Factor of rotation in CPR-W_{Ccor} plane of the compressor pressure map;

$\pi_C^*(b)$—Factor of rotation in CPR-N_{cor} plane;

$W_{HPT}(c)$—HPT flow coefficient;

$W_{PT}(c)$—Power turbine flow coefficient;

$\eta_C^*(a)$—Factor of rotation in the η_C^*–W_{Ccor} plane of the compressor efficiency map;

$\eta_C^*(c)$—Scaling factor of the compressor efficiency map;

η_{CC}—Scaling factor of combustion efficiency;

3.1. Least Squares Identification

The initial values of model parameters $\vec{\Theta}_0$ that correspond to the average engine (a priori information) are shown in Table 2.

Table 2. Initial values of the model parameters.

$\pi_{HPC}^*(a)$	$\pi_{HPC}^*(b)$	$\pi_{HPC}^*(c)$	$W_{HPT}(c)$	$\eta_{HPC}^*(a)$	$\eta_{HPC}^*(c)$	$W_{HPT}(c)$	$W_{PT}(c)$	$\eta_{PT}^*(c)$	η_{CC}
−0.0788	0.0125	0.023	0.0464	0.0383	0.0548	0.0265	−0.0938	0.01	0.02

In the first step of the iteration procedure, the following corrections to component maps were determined:

$$\delta\pi_C^*(a) = 342\ \%;\ \delta W_{HPT} = 0.93\ \%;\ \delta W_{PT} = 0.85\ \%;\ \delta\eta_C^*(a) = 26\ \%;\ \delta\eta_C^*(c) = 1.4\ \% \qquad (19)$$

Such large corrections make it impossible to calculate part-load performance and influence coefficient matrices for the next iteration. Therefore, the calculation procedure was modified: Corrections to the component maps were limited. At corrections as high as 7%, the procedure slowly becomes convergent (by 7–9 iterations). The solution corresponds to small (less than 1%) corrections to the component maps.

The results of the standard least squares (the Newton–Raphson method) show that this method has low stability due to a weak conditionality of the engine model. Therefore, practical application requires improvement of the method's stability.

3.2. Regularized Identification

The same experimental data were processed by the regularized method based on Equation (9). Parameter α was set at 0.01. Table 3 shows corrections $\delta\vec{\Theta}^s$ for the first three iterations while Table 4 presents the parameters calculated by the corrected model.

Table 3. Corrections for three first iterations.

Iteration	$\delta\pi_C^*(a)$	δW_{HPT}	δW_{PT}	$\delta\eta_C^*(a)$	$\delta\eta_C^*(a)$
1	0.004	0.17	0.39	−0.033	0.33
2	0.003	0.18	0.15	−0.018	0.08
3	0.003	0.09	0.08	−0.016	0.02

Table 4. Values of measured parameters determined by the corrected model.

Operating Point	t_H, °C	P^*_H, Pa	N_2, %	N_1, %	P_{net}, kW	W_f, kg/h	TIT, K	CPR
1st cruise	12.0	100,192	92.38	98.42	1223	374.2	1050.3	7.48
2nd cruise	11.6	100,192	94.44	98.20	1521	439.8	1105.7	8.30
Nominal	12.3	100,192	96.00	98.10	1729	490.1	1146.5	8.84
Maximum	11.9	100,192	98.86	98.12	2203	601.0	1224.2	10.18

Table 3 shows the high stability of the procedure: Estimations change slowly, without significant oscillations. The changes become smaller from iteration to iteration (this is a sign of stability).

Comparison of Tables 1 and 4 shows that parameters determined by the corrected model are well correlated with the measured parameters. Thus, the developed procedure of component performance estimation and the engine model matching can be implemented in practice.

3.3. Regularized Multi-Criteria Identification

Finally, the above task of multi-mode diagnostics was considered with a priori information about the model of the average engine, the parameters of which were estimated by previous testing data (Table 2). It is also known that a scatter of measured parameters of different engines at P_{net} = const, N_1 = const follows the normal distribution with the following mean squared error (Table 5).

Table 5. Mean squared error of performance parameters in the engine population.

Parameter	N_2	W_f	T_3	π_c
Mean squared error	0.85%	0.7%	1.1%	0.6%

This information was then transformed into the functional Φ_a (Equation (13)), and taking into account that in this case, the functionals Φ_e and Φ_a are homogeneous as they contain residuals by parameters of the same names. This facilitated the setting of weight coefficients in Equation (14). They had to relate to a scatter of measurements and a scatter of the engine parameters by series. This provided the composition of the functionals:

$$\Phi(\vec{\Theta}) = \Phi_e(\vec{\Theta}) + \frac{\sigma^2_{meas\ av}}{\sigma^2_0}\Phi_a(\vec{\Theta}) \tag{20}$$

In the considered example $\frac{\sigma^2_{meas\ av}}{\sigma^2_0} \approx 8$.

Table 6 shows the estimated parameters of the model in subsequent iterations. The final results of the corrected model are presented in Table 7. The comparison of Tables 1, 4 and 7 confirms that the proposed procedure is stable. For the corrected model, deviations of output parameters from experimental data are within the measurement error. The introduction of a priori information results in a smaller deviation of results compared to the conventional regularization method.

Table 6. Corrections for three first iterations.

Iteration	$\delta\pi_C^*(a)$	δW_{HPT}	δW_{PT}	$\delta\eta_C^*(a)$	$\delta\eta_C^*(a)$
1	0.005	0.15	0.27	−0.04	0.28
2	0.003	0.12	0.13	−0.02	0.06
3	0.002	0.07	0.06	−0.015	0.02

Table 7. Values of measured parameters determined by the corrected model.

Operating Point	t_H, °C	P^*_H, Pa	N_2, %	N_1, %	P_{net}, kW	W_f, kg/h	TIT, K	CPR
1st cruise	12.0	100,192	94.58	94.58	1223	371.7	1013	7.98
2nd cruise	11.6	100,192	96.87	96.87	1521	438.2	1066	8.85
Nominal	12.3	100,192	98.51	98.51	1729	484.0	1100.5	9.43
Maximum	11.9	100,192	102.02	102.02	2203	600.2	1186	10.65

Figure 8 presents the experimental data and modelling results for part-load performance. As the initial model is far enough from the field data, the diagnostic algorithm needs a preliminary adjustment to make the model closer to the average engine and to improve the precision of GPA. The visible lines correspond to the initial model before the correction, the average engine model and corrected model of the analyzed engine. The deviations of performance parameters from the reference indicate the engine's health status.

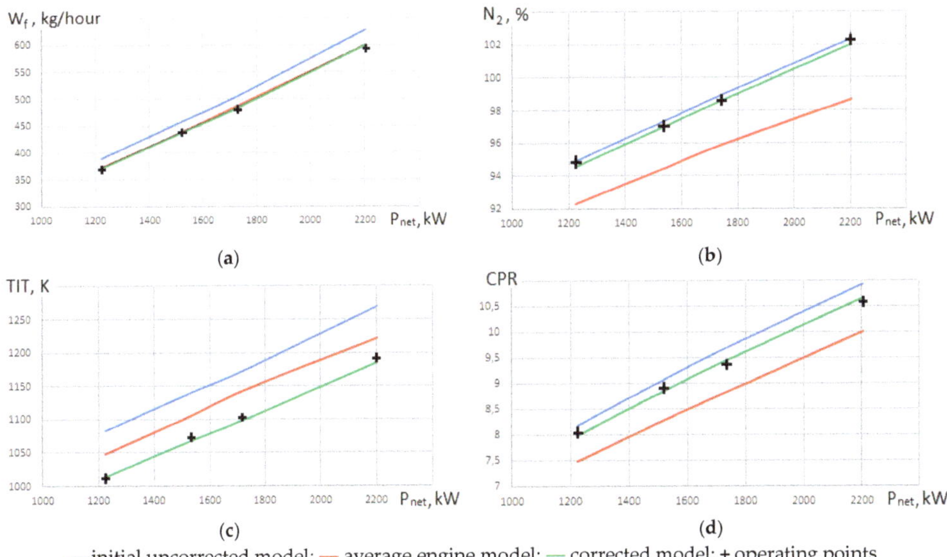

Figure 8. Experimental data vs models. (**a**) fuel flow; (**b**) high pressure rotor rotation speed; (**c**) turbine inlet temperature; (**d**) compressor pressure ratio.

Figure 9 illustrates the correction of compressor maps and confirms that the procedure was effective even though the initial model was inaccurate.

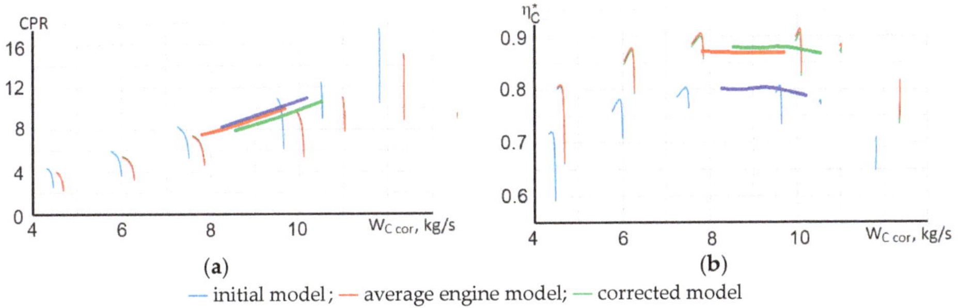

Figure 9. Compressor maps with the operating lines.

4. Discussion

Matching turbine engine models to experimental data is an inverse problem of mathematical modelling which is often characterized by parametric uncertainty. This results from the fact that the number of measured parameters is significantly lower than the number of components' performance parameters needed to describe the real engine. It was shown that in such conditions, even small measurement errors resulted in a high variation of results, hence the obtained efficiency, loss factors, etc., exceeded physically feasible ranges.

It was confirmed that extending the least squares functional by regularizing term improved stability of estimation but caused significant bias, which could lead to incorrect diagnoses. Stable operation of the algorithm and tolerant bias of the estimates in wide range of engine operation conditions are difficult to achieve as only selected values of the regularization coefficient α can ensure that. Finding its optimal solution is complicated because the regularizing term is not related to the physical properties of the object.

The new regularization procedure of multi-criteria identification, introduced in Section 2.4, uses a priori information about the engine, its measurement system, mathematical model and expected performance (Figure 1). This information is heterogeneous: Partially, it is presented in the form of statistic parameters and partially in the form of heuristic expert representations. Some other methods, e.g., Extended Kalman Filter can also use a priori information [22]. Undoubtedly, the quality of implemented expert knowledge should be carefully checked in advance.

The framework that supported the formalization of the regularization problem used fuzzy sets theory [33,37]. Similarly to the solution of Tanaka et al. [38], membership functions were introduced directly into the minimized functional, which contained a priori information about an acceptable range of estimated parameters. However, fuzzy logic is not used here for the non-linear mapping of parameters like in AI-based diagnostics [39]. In this work, inputs and outputs are measured variables with Gaussian noise.

The integral form of the stabilization functional was proposed which operates with continuous membership functions. To accelerate numerical calculations, the trapezoidal membership function was recommended. The analytical equation was derived for the stabilization functional with this function, which was used in place of numerical integration, thus decreasing the computing capacity required for the estimation procedure.

In the example presented in Section 3.3, available component maps were significantly different from the actual maps of the tested engine. Nevertheless, the estimation algorithm worked stable and provided a perfect matching of the model to the experimental data. Thanks to regularizing terms implemented in the estimation procedure, intermediate estimations did not exceed their limits when the computational algorithm was unstable.

Further studies and more field data are needed to quantify the accuracy and the stability of the procedure in relation to the number of estimated parameters. However, when a large number of engine parameters is measured, measurement uncertainty is low, and the number of coefficients of the model to be found is significantly less than the number of measured parameters, the estimates will not exceed the acceptable range, stability is not a problem, and the regularization does not have a positive effect and is not required. Hence, the proposed method is efficient when the number of measured parameters is small, and errors of measurement are significant. This is often the case in real applications.

5. Conclusions

A new method for the stable estimation of engine performance parameters is proposed. It uses a priori information about the engine and data acquisition system, i.e., the expected performance and acceptable variations of parameters. The method improves the accuracy of gas-turbine performance models in off-design conditions. A priori information is introduced in the model identification procedure in the form of membership functions, which are transformed into additional stabilization terms of the functional to be minimized.

The proposed approach was used to process test-cell data from a helicopter engine. The comparison with the conventional least squares and Tikhonov regularization proved the following:

(1) LSM has low stability that can cause the engine simulation software to crash.
(2) Conventional regularization improves the stability, but the intensive bias of the estimates needs to be excluded in advance.
(3) A priori information has a physical sense, so it improves the stability and precision of the estimation. The introduced identification method is effective in ill-conditioned configurations and provides small bias of estimates.

Author Contributions: Conceptualization, O.K. and R.Z.; methodology, O.K.; software, R.Z.; validation, R.P. and S.Y.; formal analysis, S.Y. and R.Z.; investigation, O.K.; writing—original draft preparation, S.Y.; writing—review and editing, R.P.; visualization, S.Y., R.Z. and R.P. All authors have read and agreed to the published version of the manuscript.

Funding: This publication was prepared within the framework of the AERO-UA project, which has received funding from the European Union's Horizon 2020 research and innovation programme under grant agreement No. 724034, and with the support of the Ministry of Education and Science of Ukraine (research project No. D203-3/2019-Π).

Acknowledgments: The authors would like to acknowledge the constructive comments, and suggestions provided by the anonymous reviewers that greatly improved the quality of the article. We are also grateful to Michail Ugryumov and Anatolyi Mazurkov for their help in the application of the genetic optimization procedure and Kacper Grzędziński for his comments on an earlier version of the manuscript.

Conflicts of Interest: The authors declare no conflict of interest.

Nomenclature

The following abbreviations and symbols are used in this manuscript:

AI	Artificial intelligence
av	Average value
C	Generalized matrix of influence coefficients compressor
CC	Combustion Chamber
CPR	Compressor Pressure Ratio
G	Weight diagonal matrix
GPA	Gas Path Analysis
H	Influence coefficient matrix
HPC	High Pressure Compressor
HPT	High Pressure Turbine
I	Identity matrix
ICM	Influence coefficient matrix
N	Number of operating points, rotational speed
LPT	Low Pressure Turbine
LSM	Least Squares Method
m	Number of measured parameters
OP	Operating Point
P	Pressure
P_{net}	Net thrust
r	Number of estimated parameters
sim	Simulated value
TIT	Turbine Inlet Temperature
U	Parameters that determine the engine operating conditions
W	Mass flow rate of air/gas
W_f	Fuel flow
Y	Measured parameters
Z	Generalized vector of measured parameters
α	Regularization (weighting) factor
Δ	Parameter's correction
δ	Relative deviation
η	Efficiency
σ^2	Variance
π	Pressure ratio
σ	Pressure loss coefficient
θ	Component's map parameter
$\rightarrow$	Vector
$\hat{}$	Estimated parameter
$*$	Measured value

References

1. Simon, D.L.; Borguet, S.; Léonard, O.; Zhang, X.F. Aircraft Engine Gas Path Diagnostic Methods: Public Benchmarking Results. In Proceedings of the ASME Turbo Expo 2013: Technical Conference and Exposition, San Antonio, TX, USA, 3–7 June 2013; GT2013-95077. p. 13. [CrossRef]
2. Fentaye, A.D.; Baheta, A.T.; Gilani, S.I.; Kyprianidis, K.G. A Review on Gas Turbine Gas-Path Diagnostics: State-of-the-Art Methods, Challenges and Opportunities. *Aerospace* **2019**, *6*, 83. [CrossRef]
3. Stamatis, A.; Mathioudakis, K.; Papailiou, K.D. Adaptive Simulation of Gas Turbine Performance. *J. Eng. Gas Turbines Power* **1990**, *112*, 168–175. [CrossRef]
4. Sampath, S.; Singh, R. An Integrated Fault Diagnostics Model Using Gen10etic Algorithm and Neural Networks. *J. Eng. Gas Turbines Power* **2008**, *130*, 49–56. [CrossRef]
5. Lu, F.; Chunyu, J.; Huang, J.; Wang, J.; You, C. A Novel Data Hierarchical Fusion Method for Gas Turbine Engine Performance Fault Diagnosis. *Energies* **2016**, *9*, 828. [CrossRef]
6. Bayona-Roa, C.; Solís-Chaves, J.; Bonilla, J.; Rodriguez-Melendez, A.; Castellanos, D. Computational Simulation of PT6A Gas Turbine Engine Operating with Different Blends of Biodiesel—A Transient-Response Analysis. *Energies* **2019**, *12*, 4258. [CrossRef]
7. Lambiris, B.; Mathioudakis, K.; Stamatis, A.; Papailiou, K. Adaptive Modeling of Jet Engine Performance with Application to Condition Monitoring. *J. Propuls. Power* **1994**, *10*, 890–896. [CrossRef]
8. Kong, C.; Ki, J.; Kang, M. A New Scaling Method for Component Maps of Gas Turbine Using System Identification. *J. Eng. Gas Turbines Power* **2003**, *125*, 979–985. [CrossRef]
9. Li, Y.G.; Pilidis, P.; Newby, M.A. An Adaptation Approach for Gas Turbine Design-point Performance Simulation. *J. Eng. Gas Turbines Power* **2006**, *128*, 789–795. [CrossRef]
10. Li, Y.G.; Ghafir, M.F.A.; Wang, L.; Singh, R.; Huang, K.; Feng, X. Nonlinear Multiple Points Gas Turbine Off-design Performance Adaptation Using a Genetic Algorithm. *J. Eng. Gas Turbines Power* **2011**, *133*, 071701. [CrossRef]
11. Roth, B.A.; Doel, D.L.; Cissell, J.J. Probabilistic Matching of Turbofan Engine Performance Models to Test Data. In Proceedings of the ASME Turbo Expo 2005: Power for Land, Sea and Air, Reno-Tahoe, NV, USA, 6–9 June 2005; GT2005-68201. pp. 541–548. [CrossRef]
12. Tsoutsanis, E.; Li, Y.G.; Pilidis, P.; Newby, M. Non-linear Model Calibration for Off-design Performance Prediction of Gas Turbines with Experimental Data. *Aeronaut. J.* **2017**. [CrossRef]
13. Ramadhan, E.; Li, Y.-G.; Maherdianta, D. Application of Adaptive GPA to an Industrial Gas Turbine Using Field Data. In Proceedings of the ASME Turbo Expo 2019: Turbomachinery Technical Conference and Exposition, Phoenix, AZ, USA, 17–21 June 2019; GT2019-90686. p. 11. [CrossRef]
14. Zaccaria, V.; Stenfelt, M.; Sjunnesson, A.; Hansson, A.; Kyprianidis, K.G. A Model-Based Solution for Gas Turbine Diagnostics: Simulations and Experimental Verification. In Proceedings of the ASME Turbo Expo 2019: Turbomachinery Technical Conference and Exposition, Phoenix, AZ, USA, 17–21 June 2019; GT2019-90858. p. 11. [CrossRef]
15. Litvinov, Y.; Borovick, V. *Performances and Maintenance Properties of Aircraft Turbine Engines*; Mechanical Engineering Publishing: Moscow, Russia, 1979; p. 288.
16. Yepifanov, S. Turbine Engine Parameters Optimum Estimation on Experimental Data. In Proceedings of the Scientific Conference Gas Turbine and Combined Power Plants, Moscow, Russia, 17–19 November 1983; p. 14.
17. Yepifanov, S.; Kuznetsov, B.; Bogayenko, I. *Synthesis of Turbine Engine Control and Diagnosing Systems*; Technical Publishing: Kiev, Ukraine, 1998; p. 312.
18. Vapnik, V. *Dependences Reconstruction by Empirical Data*; Science Publishing: Moscow, Russia, 1979; p. 447.
19. Volponi, A.J.; DePold, H.; Ganguli, R.; Daguang, C. The Use of Kalman Filter and Neural Network Methodologies in Gas Turbine Performance Diagnostics: A Comparative Study. *J. Eng. Gas Turbines Power* **2003**, *125*, 917–924. [CrossRef]
20. Simon, D. A Comparison of Filtering Approaches for Aircraft Engine Health Estimation. *Aerosp. Sci. Technol.* **2008**, *12*, 276–284. [CrossRef]
21. Bourguet, S.; Leonard, O. Comparison of Adaptive Filters for Gas Turbine Performance Monitoring. *J. Comput. Appl. Math.* **2010**, *204*, 2202–2212. [CrossRef]

22. Lu, F.; Ju, H.; Huang, J. An Improved Extended Kalman Filter with Inequality Constraints for Gas Turbine Engine Health Monitoring. *Aerosp. Sci. Technol.* **2016**, *58*, 36–47. [CrossRef]

23. Pourbabaee, B.; Meskin, N.; Khorasani, K. Sensor Fault Detection, Isolation, and Identification Using Multiple-model-based Hybrid Kalman Filter for Gas Turbine Engines. *IEEE Trans. Control Syst. Technol.* **2016**, *24*, 1184–1200. [CrossRef]

24. Yang, Q.; Li, S.; Cao, Y.; Gu, F.; Smith, A. A Gas Path Fault Contribution Matrix for Marine Gas Turbine Diagnosis Based on a Multiple Model Fault Detection and Isolation Approach. *Energies* **2018**, *11*, 3316. [CrossRef]

25. Chang, X.; Huang, J.; Lu, F.; Sun, H. Gas-path Health Estimation for an Aircraft Engine Based on a Sliding mode observer. *Energies* **2016**, *9*, 15. [CrossRef]

26. Chang, X.; Huang, J.; Lu, F. Robust In-flight Sensor Fault Diagnostics for Aircraft Engine Based on Sliding Mode Oobservers. *Sensors* **2017**, *17*, 835. [CrossRef]

27. Chang, X.; Huang, J.; Lu, F. Health Parameter Estimation with Second-order Sliding Mode Observer for a Turbofan Engine. *Energies* **2017**, *10*, 1040. [CrossRef]

28. Ljung, L.; Singh, R.; Chen, T. Regularization Features in the System Identification Toolbox. *IFAC-PapersOnLine* **2015**, *48*, 745–750. [CrossRef]

29. Breikin, T.V.; Arkov, V.Y.; Kulikov, G.G. Regularization Approach for Real-time Modelling of Aero Gas Turbines. *Control Eng. Pract.* **2004**, *12*, 401–407. [CrossRef]

30. Guseynov, S.E.; Aleksejeva, J.V.; Andreyev, S.A. On one Regularizing Algorithm for Comprehensive Diagnosing of Apparatus, Engines and Machinery. *Adv. Mater. Res.* **2015**, *1117*, 254–257. [CrossRef]

31. Guseynov, S.E.; Yunusov, S.M. New Regularizing Approach to Determining the Influence Coefficient Matrix for Gas-turbine Engines. *Discret. Contin. Dyn. Syst. Ser. A* **2011**, 614–623. [CrossRef]

32. Roth, B.; Doel, D.L.; Mavris, D.; Beeson, D. High-Accuracy Matching of Engine Performance Models to Test Data. In Proceedings of the ASME Turbo Expo 2003: Power for Land, Sea and Air, Atlanta, GA, USA, 16–19 June 2003; GT2003-38784. p. 9. [CrossRef]

33. Zadeh, L.A. The Linguistic Approach and its Application to Decision Analysis. In *Advances in Fuzzy Systems—Applications and Theory, Fuzzy Sets, Fuzzy Logic, and Fuzzy Systems*; Directions in large-scale systems; World Scientific: Singapore, 1996; pp. 260–282. [CrossRef]

34. Zwingenberg, M.; Benra, F.-K.; Werner, K.; Dobrzynski, B. Generation of Turbine Maps Using a Fusion of Validated Operational Data and Streamline Curvature Method. In Proceedings of the ASME Turbo Expo 2010: Power for Land, Sea, and Air, Volume 7: Turbomachinery, Parts A, B, and C, Glasgow, UK, 14–18 June 2010; pp. 2757–2768. [CrossRef]

35. Meniailov, I.; Khustochka, O.; Ugryumova, K.; Chernysh, S.; Yepifanov, S.; Ugryumov, M. Mathematical Models and Methods of Effective Estimation in Multi-objective Optimization Problems Under Uncertainties. In *Advances in Structural and Multidisciplinary Optimization*; Springer: Cham, Switzerland, 2018; pp. 411–427. [CrossRef]

36. Zhang, Y.; Martínez-García, M.; Kalawsky, R.S.; Latimer, A. Grey-box modelling of the swirl characteristics in gas turbine combustion system. *Measurement* **2020**, *151*, 107266. [CrossRef]

37. Zhang, Y.; Martinez-Garcia, M.; Serrano-Cruz, J.R.; Latimer, A. Multi-region System Modelling by using Genetic Programming to Extract Rule Consequent Functions in a TSK Fuzzy System. In Proceedings of the 4th IEEE International Conference on Advanced Robotics and Mechatronics, ICARM 2019, Toyonaka, Japan, 3–5 July 2019; pp. 987–992. [CrossRef]

38. Tanaka, H.; Uejima, S.; Asai, K. Linear Regression Analysis with Fuzzy Model. *IEEE Trans. Syst. Man Cybern.* **1982**, *12*, 903–907. [CrossRef]

39. Cao, Y.; Hu, P.; Yang, Q.; He, Y.; Li, S.; Yu, F.; Du, J. Fuzzy Analytic Hierarchy Process Evaluation Method of Gas Turbine Based on Health Degree. In Proceedings of the ASME Turbo Expo 2018: Turbomachinery Technical Conference and Exposition. Volume 6: Ceramics; Controls, Diagnostics, and Instrumentation; Education; Manufacturing Materials and Metallurgy, Oslo, Norway, 11–15 June 2018. V006T05A012. [CrossRef]

Article

Superhydrophobic Coatings as Anti-Icing Systems for Small Aircraft

Filomena Piscitelli [1,*], Antonio Chiariello [1], Dariusz Dabkowski [2], Gianluca Corraro [1], Francesco Marra [3] and Luigi Di Palma [1]

[1] Department of Materials and Structures and Department of On Board Systems, CIRA–Italian Aerospace Research Centre, Via Maiorise, 1, 81043 Capua, Italy; a.chiariello@cira.it (A.C.); g.corraro@cira.it (G.C.); L.DiPalma@cira.it (L.D.P.)

[2] PW-Metrol, Książnice 28, 39-300 Mielec, Poland; p.w.metrol@wp.pl

[3] Department of Chemical Engineering, Materials and Environment, Sapienza University of Rome, Italy–INSTM Reference Laboratory for Engineering of Surface Treatments, 00185 Rome, Italy; francesco.marra@uniroma1.it

* Correspondence: f.piscitelli@cira.it

Received: 26 November 2019; Accepted: 25 December 2019; Published: 2 January 2020

Abstract: Traditional anti-icing/de-icing systems, i.e., thermal and pneumatic, in most cases require a power consumption not always allowable in small aircraft. Therefore, the use of passive systems, able to delay the ice formation, or reduce the ice adhesion strength once formed, with no additional energy consumption, can be considered as the most promising solution to solve the problem of the ice formation, most of all, for small aircraft. In some cases, the combination of a traditional icing protection system (electrical, pneumatic, and thermal) and the passive coatings can be considered as a strategic instrument to reduce the energy consumption. The effort of the present work was to develop a superhydrophobic coating, able to reduce the surface free energy (SFE) and the work of adhesion (W_A) of substrates, by a simplified and non-expensive method. The developed coating, applied as a common paint with an aerograph, is able to reduce the SFE of substrates by 99% and the W_A by 94%. The effects of both chemistry and surface morphology on the wettability of surfaces were also studied. In the reference samples, the higher the roughness, the lower the SFE and the W_A. In coated samples with roughness ranging from 0.4 and 3 μm no relevant variations in water contact angle, nor in SFE and W_A were observed.

Keywords: superhydrophobic; coating; anti-icing; spray-coat; aeronautical

---

1. Introduction

Typically, the presence of tiny pieces of ice or supercooled liquid water in the clouds, which remain liquid below zero and suddenly turn to ice after the impact with the aircraft surfaces, are the main sources of ice deposition during a flight. The presence of ice on surfaces alters the airflow over the wing and tail, and then reduces the lift force that keeps the plane in the air. This potentially causes aerodynamic stall, a condition that can lead to a temporary loss of control of the aircraft. In order to prevent or reduce the ice formation or alternatively to remove the ice once it was formed, anti-icing and de-icing systems are usually adopted. Currently the thermal and pneumatic types represent the most largely employed systems. In details, the thermal system melts the ice accretion or prevents the ice from forming by the application of heat on the protected surface of the wing. The heat is generated by the hot air "bled" off the jet engine into piccolo tubes routed through wings, tail surfaces, and engine inlets. The spent bleed air is then exhausted through holes in the lower surface of the wing. Similarly, the electro-thermal systems use resistive circuits buried in the airframe structure to generate heat when a current is applied. The heat can be generated continuously to protect the aircraft from icing (anti-ice mode), or intermittently to shed ice as it accretes on key surfaces (de-ice), the latter being generally preferred due to the lower power consumption. Instead, the pneumatic de-ice boot consists of a rubber sheet bonded to the leading edge of the airfoil: when the ice builds up on the leading edge, an engine-driven pneumatic pump inflates the rubber boots, and then the ice is cracked and should fall off the leading edge of the wing. Additionally, some innovative systems are being developed, e.g., the electro-mechanical de-icing systems which use a mechanical force, generated for instance by actuators, to knock the ice off the flight surface, or the weeping wing system, which releases a glycol-based chemical onto the wing surface through small orifices which causes the detachment of ice. However, these methods can be inefficient, environmentally unfavorable, expensive and time consuming [1]. Thus, it would be advantageous if surfaces could passively prevent the ice formation and facilitate ice removal. In this contest, the superhydrophobic coatings, applied as usual paint, owing to their extraordinary water repellency, and not requiring additional energy consumption, can be viewed as excellent candidates for icephobicity in this area [2,3]. This is because, ultrahigh θ_{rec}, typical of nanostructured superhydrophobic surfaces, means low ice adhesion strength [4,5]. Subsequently, if supercooled liquid water freezes when it impacts the solid surface of an aircraft, the resulting ice can be shed, taking advantage of external forces, such as aerodynamic forces, to overcome ice-surface adhesion forces [6]. Evidently, the efficiency of these coatings as an anti-icing passive method largely depends on the different scenarios in which the ice may form [1]. In some cases, therefore, a combination of a traditional icing protection system and the superhydrophobic coatings could be seen as a strategic instrument able to assure a high efficiency in a wide range of environmental conditions.

In the last decade, a huge body of literature concerning the development of superhydrophobic and icephobic coatings able to delay the ice formation [7], or reduce the ice adhesion strength once formed, has been produced [8–15]. A shared strategy is to create surfaces with low surface free energy and, contemporary, micro-nanoscopic rough structures. In this regard, it was found that the icing probability was reduced to zero with particle diameters ranging between about 30 and 70 nm [16]. Some authors achieved this goal by using etching and/or high temperatures [9,11], or complex fabrication techniques [12–17], which are often limited by the substrate type and geometry that can be successfully coated [18].

The effort of the present work was to develop a superhydrophobic coating for metallic substrates with a simplified and non-expensive method, which could be employed as a usual paint able to prevent/reduce the formation of ice, especially on small aircraft [19]. The newly formulated superhydrophobic coating consists of nanostructured layers able to generate hierarchical micro/nano-structured roughness, and reduce the surface free energy, which as previously declared, are the two main factors useful to making a superhydrophobic surface. The effect of the substrates' roughness on the surface properties of coated and uncoated samples was also studied, and then correlated to the wettability of samples, in order to identify and separate the morphological and the chemical contributions.

This work is being developed in the framework of the Clean Sky 2 ongoing SAT-AM (More Affordable Small Aircraft Manufacturing) project, whose main goal is the investigation of new technologies for a future small aircraft able to fly with low fuel consumption, low noise levels and needing low quantities of raw material in its life cycle. The reference vehicle for this project is the M28 designed and manufactured by Consortium Partner Polskie Zakłady Lotnicze (PZL), Mielec (PL). It is a commuter category 19 passenger, twin-engine high-wing cantilever monoplane of all-metal structure, with twin vertical tails and a robust tricycle non-retractable landing gear, featuring a steerable nose wheel to provide for operation from short, unprepared runways where hot or high-altitude conditions may exist. The M28 is best suited for passenger and/or cargo transportation and is certified under EU CS-23 and USA FAR 23 requirements.

2. Materials and Methods

The developed superhydrophobic coating was prepared starting from the formulation described in a previous work [19] and slightly modified in order to improve its durability. In this regard, other details about the performed changes cannot be disseminated in this manuscript. Once prepared, several layers of the developed coating were applied with an aerograph on substrates representative of the M28 air intake. The dehumidified air was pressured at three bar; whereas the other application parameters, such as the layers' number, the curing temperature and the distance between the aerograph and samples were optimized in order to guarantee uniformity in the coating's thickness and weight.

The substrate samples, 5×5 cm^2 in dimensions, made of stainless steel 12H18N10T (same composition of the M28 air intake), and having roughness ranging from 0.7 to 4.6 μm, were provided by METROL (a partner of the Clean Sky project). The effect of the substrate roughness on the final properties of the coating in terms of wettability and adhesion of the coating on the substrate were studied.

Roughness of substrates before and after the application of the coating was measured by employing a SAMA SA6260 surface roughness meter. Roughness measures, performed according to the ISO 4288 [20], were reported as Ra, which represents the arithmetic average of the absolute values of the profile height deviations from the mean line. ID samples was correlated to the correspondent roughness values in Table 1.

Table 1. ID samples with roughness ranging between 0.7 and 4.6 μm.

Sample ID	Roughness of Uncoated Samples [μm]
1	0.752
2	0.83
3	0.91
4	1.241
5	1.341
6	1.541
7	4.032
8	4.360
9	4.623

Contact angles (CA) were calculated according to Young's equation [21] (see Figure 1):

$$\gamma_{LV} \cos \theta = \gamma_{SV} - \gamma_{SL} \tag{1}$$

where θ is the contact angle, γ_{LV} and γ_{SV} are the liquid and solid surface free energy, respectively, whereas γ_{SL} is the solid/liquid interfacial free energy. The schematic illustration of the equilibrium among involved forces according to the Young equation is shown in Figure 1.

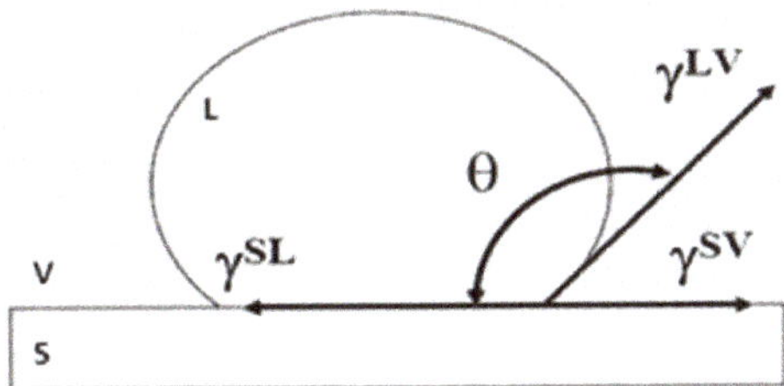

Figure 1. Schematic illustration of an ink drop on a solid substrate with the surface free energy and the contact angle as described by Young's equation [21].

The contact angle measurements were performed at room temperature in compliance with the ASTM D7490–13 [22] standard, using water (H_2O) and diiodomethane (CH_2I_2). The Surface Free Energy (SFE) was calculated according to the Owens Wendt (OW) method [23,24], for which the surface energy of a solid is the sum of two components, a dispersive and a polar one. The dispersive component theoretically accounts for the Van der Waals and other non-site specific interactions between the surface and the applied liquid, whereas the polar component accounts for the dipole-dipole, dipole-induced dipole, hydrogen bonding, and other site-specific interactions [25]. The polar and dispersive components of the reference liquids are listed in Table 2. Tests were carried out by depositing ten drops of 3 µL of each liquid on the sample surface.

Table 2. Surface tension components of the reference liquids [26].

Liquid	Formula	Room T Surface Tension [mN/m]	Dispersive Component [mN/m]	Polar Component [mN/m]
Water	H_2O	72.8	26.4	46.4
Diiodomethane	CH_2I_2	50.8	50.8	0
Formamide	$HCONH_2$	57.0	22.4	34.6

According to the OW method [23], the interfacial solid/liquid energy can be evaluated as:

$$\gamma_{sl} = \gamma_s + \gamma_l - 2\left(\gamma_l^d \gamma_s^d\right)^{1/2} - 2\left(\gamma_l^p \gamma_s^p\right)^{1/2} \tag{2}$$

which, combined with the Young equations, gives the following:

$$\gamma_l(1 + \cos\theta) = 2\left(\gamma_{l_1}^d \gamma_s^d\right)^{1/2} + 2\left(\gamma_{l_1}^p \gamma_s^p\right)^{1/2} \tag{3}$$

Therefore, the polar (γ_s^p) and dispersive (γ_s^d) components of the solid SFE are given as solution of the following non-linear system:

$$\begin{cases} \dfrac{\gamma_{l_1}(1+\cos\theta_1)}{2} = \left(\gamma_{l_1}^d \gamma_s^d\right)^{1/2} + \left(\gamma_{l_1}^p \gamma_s^p\right)^{1/2} \\[2mm] \dfrac{\gamma_{l_2}(1+\cos\theta_2)}{2} = \left(\gamma_{l_2}^d \gamma_s^d\right)^{1/2} + \left(\gamma_{l_2}^p \gamma_s^p\right)^{1/2} \end{cases} \tag{4}$$

where $\gamma_{l_1}^d$, $\gamma_{l_1}^p$, $\gamma_{l_2}^d$, $\gamma_{l_2}^p$ and θ_1, θ_2 are the SFE components of the two reference liquids and the corresponding contact angles between the solid and the liquids, respectively.

Although the SFE's components can be known through the literature (Table 2), the contact angles have to be experimentally evaluated through several measures. That determines a stochastic nature of the contact angles and therefore of the solutions of the equation system (4).

So, to assess the SFE problem, a third reference liquid, i.e., the Formamide ($HCONH_2$), was introduced in order to reformulate the initial problem in terms of the following nonlinear programming problem:

$$\min_{\gamma_s^d, \gamma_s^p}\left(\left(\gamma_{l_3}^d \gamma_s^d\right)^{1/2} + \left(\gamma_{l_3}^p \gamma_s^p\right)^{1/2} - \frac{\gamma_{l_3}(1+\cos\theta_3)}{2}\right) \tag{5}$$

$$s.t.\begin{cases} \left(\gamma_{l_1}^d \gamma_s^d\right)^{1/2} + \left(\gamma_{l_1}^p \gamma_s^p\right)^{1/2} - \dfrac{\gamma_{l_1}(1+\cos\theta_1)}{2} = 0 \\[2mm] \left(\gamma_{l_2}^d \gamma_s^d\right)^{1/2} + \left(\gamma_{l_2}^p \gamma_s^p\right)^{1/2} - \dfrac{\gamma_{l_2}(1+\cos\theta_2)}{2} = 0 \\[2mm] E[\theta_1] - \sigma_{\theta_1} \leq \theta_1 \leq E[\theta_1] + \sigma_{\theta_1} \\[1mm] E[\theta_2] - \sigma_{\theta_2} \leq \theta_2 \leq E[\theta_2] + \sigma_{\theta_2} \\[1mm] E[\theta_3] - \sigma_{\theta_3} \leq \theta_3 \leq E[\theta_3] + \sigma_{\theta_3} \end{cases} \tag{5a}$$

where a Gaussian distribution of the contact angles was assumed in order to statistically characterize them through the mean value ($E[\theta_1]$, $E[\theta_2]$, $E[\theta_3]$) and the standard deviation ($\sigma_{\theta_1}, \sigma_{\theta_2}, \sigma_{\theta_3}$).

The stated optimization problem allowed determining the $\overline{\theta}_1, \overline{\theta}_2, \overline{\theta}_3$ contact angles that identify the SFE components, which minimize the third Equation (5), taking into account the measure variability and the constitutive Equations (4) of the considered liquids.

Applying the described method to the reference liquids reported in Table 2, the constraints (5a) were linearized simplifying the complexity of the problem that has been solved in Matlab/Simulink environment. The developed Matlab tool provides (as output) the values of the three best contact angles $\overline{\theta}_1, \overline{\theta}_2, \overline{\theta}_3$, along with the values of the polar and dispersive components of the solid's SFE.

As an additional crosscheck, the determined optimal contact angles, namely, $\overline{\theta}_1, \overline{\theta}_2$, instead of the mean values of the experimentally measured angles measured with Water and Diiodomethane, were introduced in the widespread linearized equations [19–26]:

$$
\begin{cases}
\dfrac{\gamma_{l1}\left(1+\cos\overline{\theta}_1\right)}{2\sqrt{\gamma_{l1}^d}} = \dfrac{\sqrt{\gamma_{l1}^p}}{\sqrt{\gamma_{l1}^d}} \cdot \sqrt{\gamma_s^p} + \sqrt{\gamma_s^d} \\[3mm]
\dfrac{\gamma_{l2}\left(1+\cos\overline{\theta}_2\right)}{2\sqrt{\gamma_{l2}^d}} = \dfrac{\sqrt{\gamma_{l2}^p}}{\sqrt{\gamma_{l2}^d}} \cdot \sqrt{\gamma_s^p} + \sqrt{\gamma_s^d}
\end{cases}
\tag{6}
$$

and the required polar and dispersive components of the solid SFE, were obtained as the graphic solution of the equations system (6) through the intercept point between the two straight line (r_1 and r_2) identified by the following angular coefficients and points:

$$
r_1: \quad m_1 = \frac{\sqrt{\gamma_{l1}^p}}{\sqrt{\gamma_{l1}^d}}; \quad P_1 = \left(0 \quad ; \quad \frac{\gamma_{l1}\left(1+\cos\overline{\theta}_1\right)}{2\sqrt{\gamma_{l1}^d}}\right)
\tag{7a}
$$

$$
r_2: \quad m_2 = \frac{\sqrt{\gamma_{l2}^p}}{\sqrt{\gamma_{l2}^d}}; \quad P_2 = \left(0 \quad ; \quad \frac{\gamma_{l2}\left(1+\cos\overline{\theta}_2\right)}{2\sqrt{\gamma_{l2}^d}}\right)
\tag{7b}
$$

Once calculated the polar and dispersive components, the required solid SFE was assessed, as:

$$
\gamma_s = \gamma_s^p + \gamma_s^d
\tag{8}
$$

whereas the Work of Adhesion (W_A) can be calculated as:

$$
W_A = \gamma_l(1+\cos\theta)
\tag{9}
$$

or equally as:

$$
W_A = 2\left(\gamma_s^p\gamma_l^p\right)^{1/2} + 2\left(\gamma_s^d\gamma_l^d\right)^{1/2}
\tag{10}
$$

Finally, the Surface Polarity (SP) was calculated as:

$$
SP = \frac{\gamma^p}{\gamma^p + \gamma^d}
\tag{11}
$$

The Young Equation (1) is valid for smooth surfaces or substrate having very low roughness, but when the roughness is too high to be neglected, the equation should be corrected taking into account the actual surface morphology. Two different models can be useful to describe the wetting of textured surfaces, i.e., the Wenzel and Cassie–Baxter (Figure 2a,b respectively) models. In details, the Wenzel model [27,28] proposed a correction factor "*r*" for contact angle on rough surfaces, which is equal to the ratio of rough interfacial area over flat interfacial area under the droplet. Substituting the roughness ratio factor "*r*" in the Young's Equation (1), one can obtain the Wenzel's equation [28], i.e.,

$$\cos \theta* = r \cos \theta \tag{12}$$

where θ^* and θ are the contact angles of a droplet on a rough surface and on the corresponding smooth surface, respectively. Wenzel's model assumes no air-trapping under droplet, which may not necessary be true.

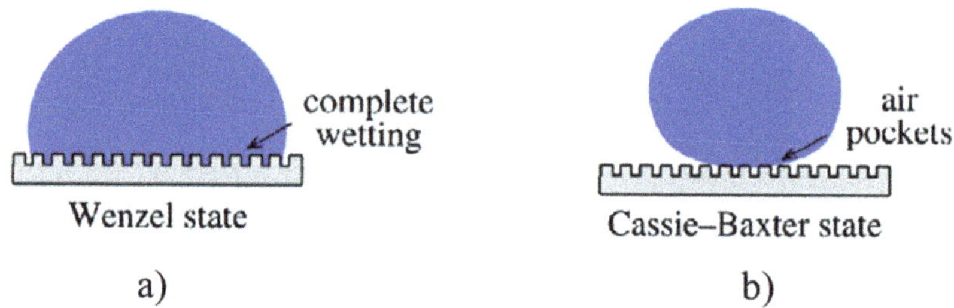

Figure 2. (**a**) Wenzel and (**b**) Cassie–Baxter models useful to describe the wetting of textured surfaces.

On the contrary, the Cassie–Baxter model [29] allows to estimating the contact angle on rough surface with air-trapping, through the following equation:

$$\cos \theta* = -1 + \varphi_S(1 + \cos \theta) \tag{13}$$

where φ_S is the area ratio of liquid-air interface to the whole interface; θ^* and θ are the contact angles with and without considering air-trapping.

Cutting and tape test were carried out, according to the ASTM D 3359-09 standard [30], by making a grid incision with a specific cutter on the coated samples and then by applying an adhesive tape to cover the cut area. The test adhesive was vigorously removed and the involved area was inspected. According to the ASTM D 3359-09 [29], the test passes if the area involved in detached flakes of coating at intersections is less than 15%.

Sample morphology, measurements of nanoparticle size and of coating thickness, were carried out using a field emission SEM Tescan Mira3 (Tescan Orsay Holding, Brno-Kohoutovice, Czech). Samples were observed in as-realized conditions, without metallization on the observed surface.

Surface 3D morphology and the corresponding roughness were assessed by using the laser profilometer Ametek Talyscan 150. The scanning area for 3D surface reconstruction was 40×40 mm^2 (800 profiles with 5 microns resolution), acquired with a scanning velocity of 10,500 microns/s.

3. Results and Discussion

3.1. Substrates' Roughness of Substrates

Roughness of substrates measured with the SAMA SA6260 surface roughness meter, before and after the application of the coating, were compared in Figure 3. It highlights that, in general, the application of the coating reduces the original roughness. This especially for samples 1–6, for which the values of the coated samples' roughness vary in the range 0.4–0.6 μm, in spite of the substrates' roughness ranging from 0.7 to 1.5 μm, whereas, for samples 7–9, the application of the coating reduces the original roughness from 4–4.5 to about 3 μm.

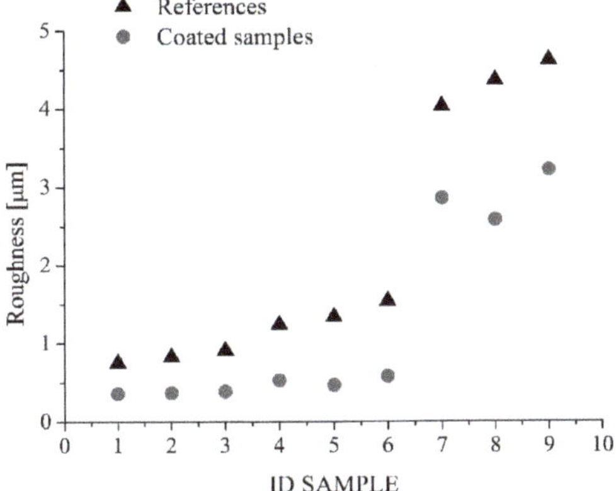

Figure 3. Roughness values of samples before (triangles) and after (circles) the application of the coating.

3.2. Application of Coating

The aerograph setting parameters were varied and optimized in order to guarantee uniformity in terms of thickness and weight of the applied coating. Here, as an example, two different setting parameters, labelled as *Procedure 1* and *Procedure 2*, are described; the corresponding schematics are shown in Figure 4. They differ mostly in the samples orientation with respect to the aerograph spray and in the drying temperature, whereas the layers' number, the coating formulation, and the substrates were the same. After the application, measures of contact angle with water at room temperature were carried out. It was found that Procedure 1 is able to produce samples with contact angle of 124°, whereas Procedure 2 gives surfaces with a water contact angle of 155° (Figure 5). This difference is probably due to the different morphologies because of the diverse application procedures.

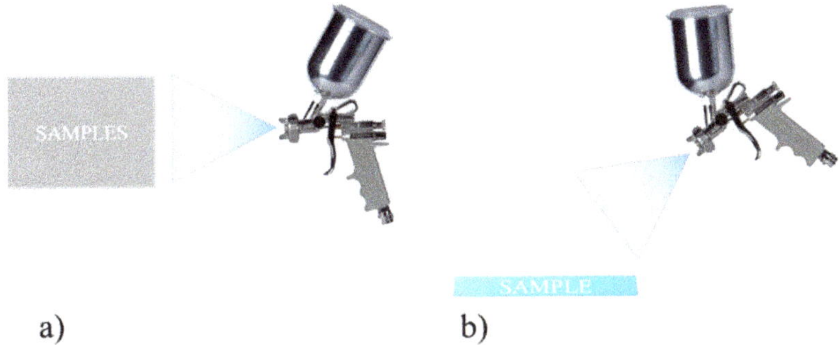

Figure 4. Schematic drawing of deposition process according to (**a**) Procedure 1 and (**b**) Procedure 2.

Procedure 1

- Vertical samples
- Drying at room temperature
- Distance between samples and aerograph = 10 cm
- P = 3 bar; hole size = 1.5 mm

Procedure 2

- Horizontal Samples
- Aerograph at 45° wrt the samples

- Drying at 75 °C of each layer of coating for 10 min
- Distance between samples and aerograph = 15 cm
- P = 3 bar; hole size = 1.5 mm;

Hence, *Procedure 2* was adopted to apply three layers of the developed coating on sample 1–9 with dimensions of 5 × 5 cm^2. The Samples were weighed before and after the application of the coating, in order to assess a rough estimation of the coating's specific weight, which was plotted as a function of the sample roughness in Figure 6. It was found that the coating's specific weight was about 11 ± 1 g/m^2.

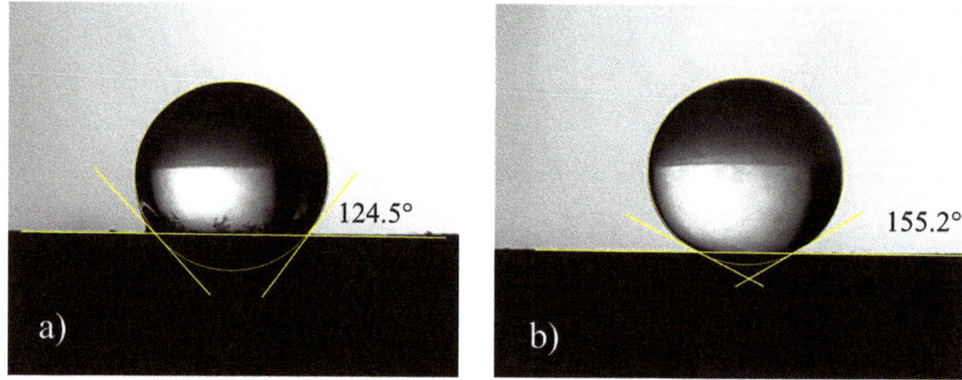

Figure 5. Pictures related to the water contact angle of coated samples prepared according to (**a**) Procedure 1 and (**b**) Procedure 2.

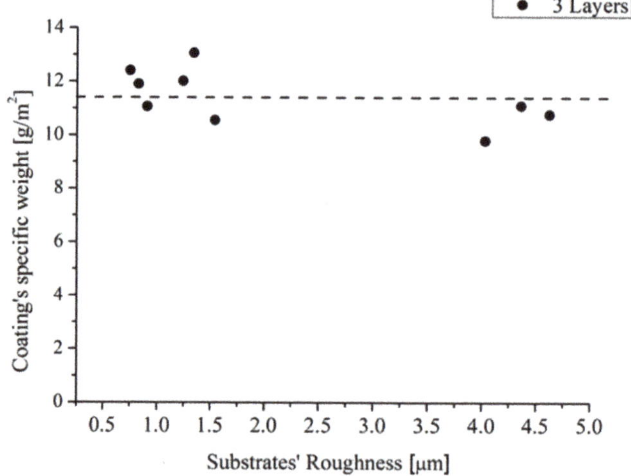

Figure 6. The coating's specific weight as a function of the substrates' roughness.

SEM cross-section images of sample 2 (Ra = 0.8 μm) and sample 7 (Ra = 4 μm), employed to measure the coating's thickness, are shown in Figure 7a,b, respectively. It seems that the two samples display some differences in both coating thickness and morphology. Sample 7 in Figure 7b, in fact, has a coating thickness higher than sample 2 shown in Figure 7b (see values in Table 3), with a mean value of 190 μm vs. 127 μm. Moreover, sample 7 coating appears to be more uniform in thickness and morphology than the sample 2 coating, and with less amount of large and localized voids. Hence, it seems that the roughness is able to improve the coating's uniformity.

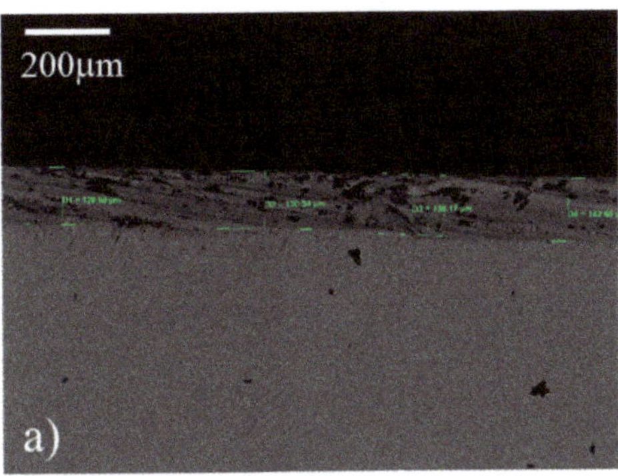
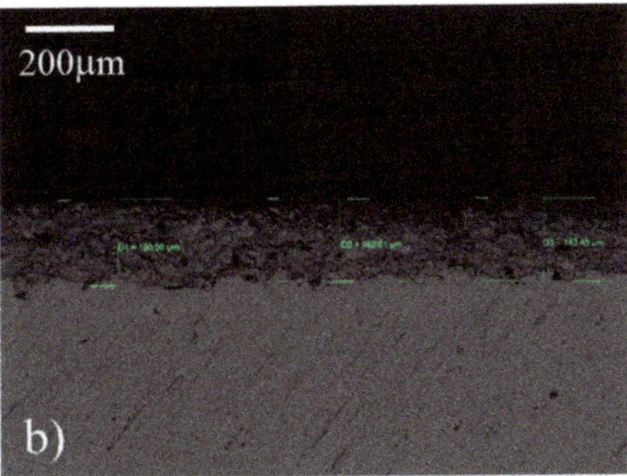

Figure 7. SEM images acquired in cross-section mode of samples 2 (**a**) and 7 (**b**). Some measured thicknesses are reported.

Table 3. Coating's thickness measured by the SEM acquired in cross section mode.

Sample ID	Coating's Thickness [µm]								Mean Value [µm]
2	112.80	115.80	123.80	125.42	129.59	130.34	136.17	142.65	127 ± 10
7	199.08	191.92	194.11	190.96	182.01	183.45			190 ± 6

3.3. Contact Angle Measurements

The measures of the contact angles acquired with H_2O and CH_2I_2 were reported as function of the substrate roughness in Figures 8 and 9, respectively. Some representative drop pictures caught during the measures were overlapped. It should be noted that, for the reference samples, the water contact angle is almost constant with a substrate roughness up to 1.8 µm beyond that, as expected [31], it increases as the roughness increases, reaching a value of about 100–115° for the highest value of measured roughness, i.e., 4.6 µm. The increasing of the contact angles with the roughness of metallic substrates can be ascribed to the changes in morphology. Other authors [31,32] also observed this behavior with maximum values of the achieved contact angles at about 120 °C [31]. Indeed, it was found that in order to achieve higher values of contact angles without modifying the surface chemistry, it is necessary to scale down the surface roughness into the sub-micron range, by creating specific and regular pillar-like structures having circular, square, triangle, crossed or fractal-like shapes [33]. On the contrary, the combination of substrate roughness and low SFE can be seen as a synergic approach able to achieve contact angle mean value of 158° (Figure 8). Additionally, it was also found that both the original and the actual sample roughness (Figures 3 and 8) do not affect the achieved contact angle, since, in spite of the roughness value, the contact angle value does not change.

The same uniformity in the achieved values of the contact angles was observed for measurements performed with CH_2I_2 in coated samples (Figure 9). Here, the mean value of the contact angles in coated samples is 148°, regardless of the substrates (0.7–4.6 µm) and the actual roughness (0.3–3.2 µm) of the samples.

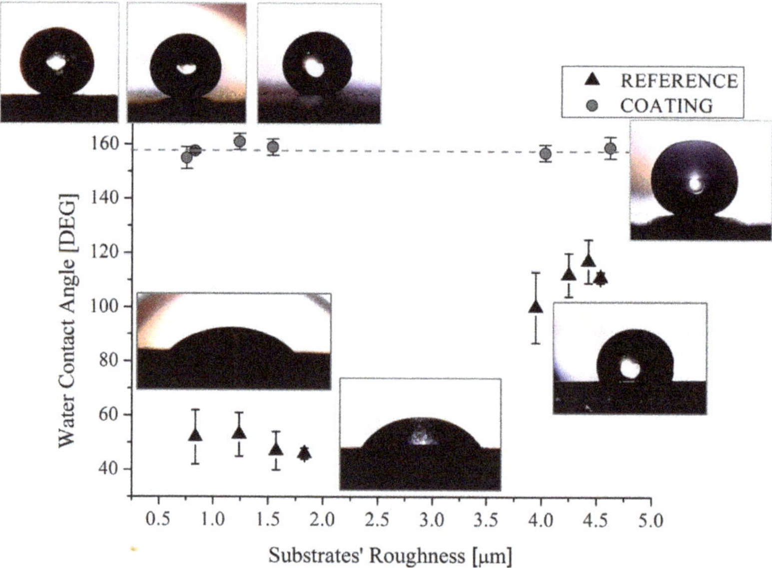

Figure 8. Water contact angles of reference and coated samples as a function of the substrates' roughness. Corresponding pictures are also shown.

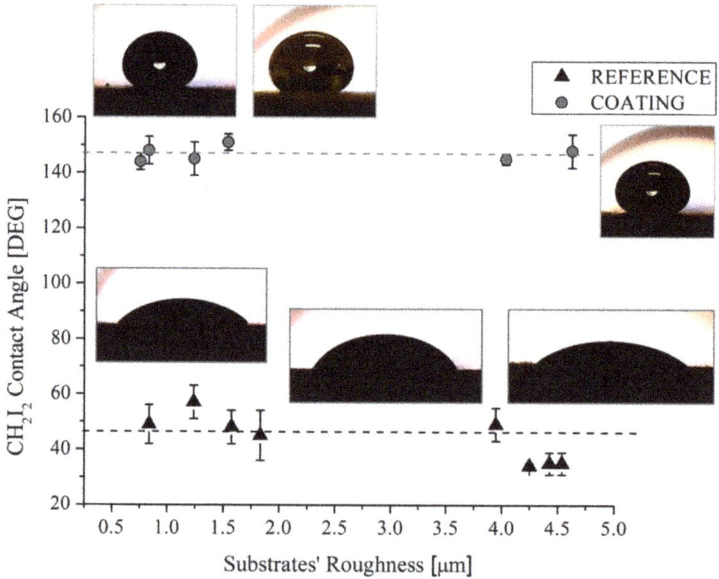

Figure 9. CH_2I_2 contact angles of reference and coated samples as a function of the substrates' roughness. Corresponding pictures are also shown.

3.4. Surface Free Energy and Work of Adhesion

The solid SFE and the W_A of both uncoated and coated samples were plotted in Figure 10a,b, respectively, as a function of the substrates' roughness.

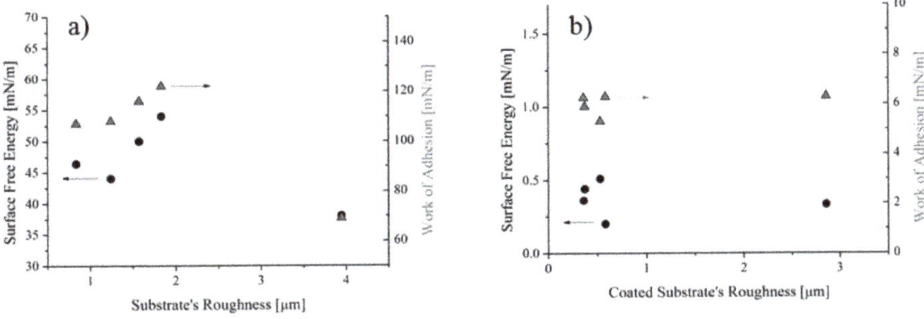

Figure 10. SFE and WA of the uncoated (**a**) and coated (**b**) samples as a function of the roughness.

It was found that the solid SFE of the reference samples varies between 44 and 54 mN/m, for roughness values ranging from 0.7 to 2 µm, whereas for roughness of 4 µm the SFE becomes 38 mN/m (Figure 10a). Similarly, the W_A varies between 108 and 122 mN/m, assuming a value of 69 mN/m for the sample at 4 µm in roughness. For samples having roughness higher than 4 µm, results cannot be achieved without taking into account the actual surface roughness, e.g., through Equations (12) and (13), or more complex relationship, i.e., those described in ref. [33]. Indeed, it was found that roughness higher than 4 µm the Equations' system (4) did not give a solution, so no intersection of the two curves can be observed in graphs of Figure 11b. On the contrary, for lower values of the contact angles, the Equation system (4) reached solutions, as shown in Figure 11a as an example.

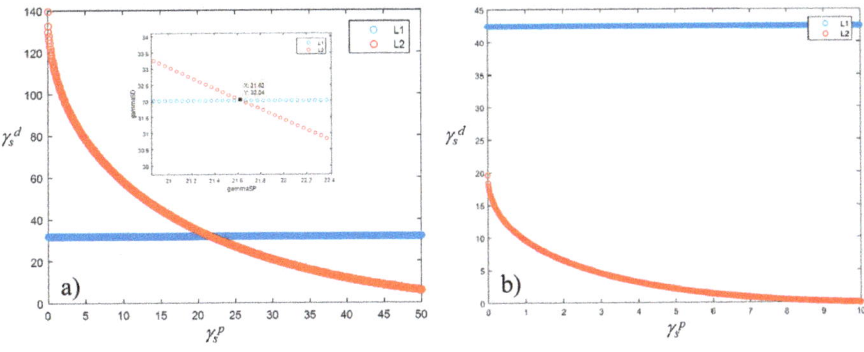

Figure 11. Graphical solution of the SFE equations' system with (**a**) and without (**b**) solution.

The authors decided not to further investigate samples at high values of roughness, since values higher than 4 µm are out of interest for aeronautical applications.

On the other hand, the flat trend for SFE and WA was seen for coated samples (Figure 10b), since all values scatter around 0.38 mN/m for the SFE and around 6.1 mN/m for the WA. This result reflects the uniformity of the water contact angle values measured in spite of the original and actual roughness values: the latter changing between 0.3 and 3 m (Figure 3). These trends can be better highlighted if results are reported as a function of the solid SFE (Figure 12). For the reference samples (Figure 12a), it was found that, as expected, the higher the contact angle, the lower the SFE and the WA. Since this result was achieved by varying the roughness of substrates alone, it can be concluded that it can be ascribed to the surface morphology (MORPHOLOGICAL EFFECT). On the contrary, for coated samples, no changes in terms of water contact angle and WA, along with almost no variations in the solid SFE (Figure 12b) can be observed, in spite of the actual roughness of substrates. Therefore, it should be concluded that, for the analyzed samples, the chemical modification of the surfaces prevails on the morphological aspect (CHEMICAL EFFECT).

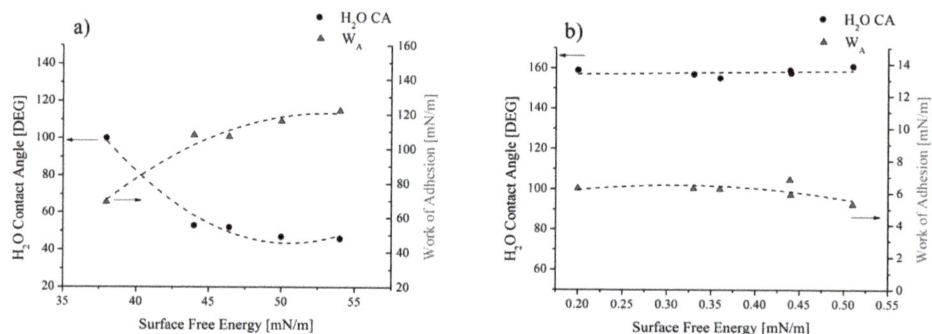

Figure 12. Water contact angle and W_A as a function of (**a**) the surface free energy for the reference and (**b**) the coated samples.

The geometrical and chemical effects on the wettability of surfaces were also observed by Kam et al. [31]. They found that the maximum contact angle achieved with nano-microstructured metallic surfaces was 110°, whereas the combination of the surface geometry and its chemical modification with silanes allowed to additionally modify the wettability of surfaces, since the contact angles reached values higher than 140°.

Finally, in Figure 13 the polar and dispersive components of the surface free energy, and the WA assessed for the reference and the coated samples were compared. It highlights that the WA was reduced by 94%, the solid SFE by 99%, and the SP by 100%, since after the application of the coating, it changes from 34 to 0%.

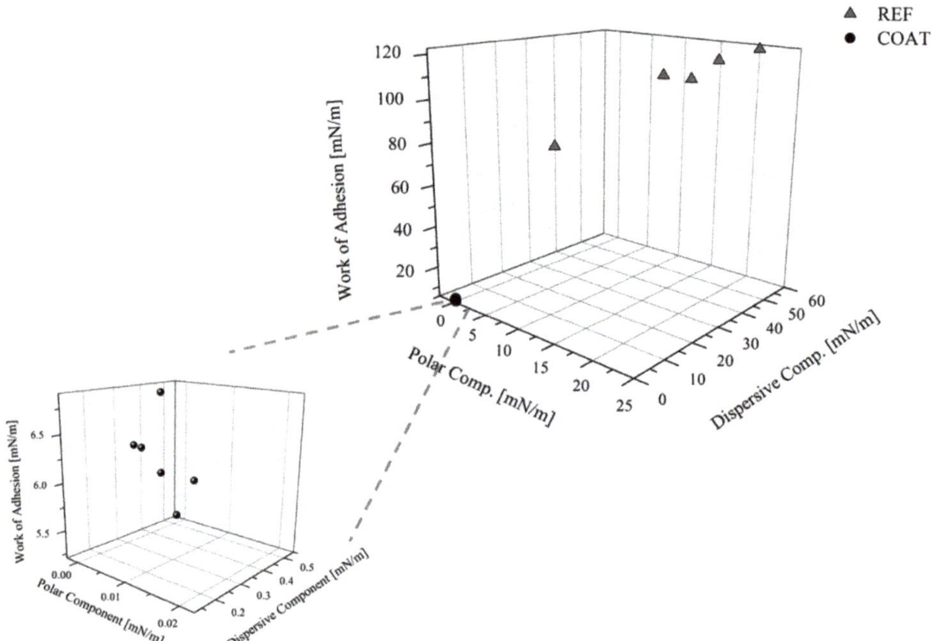

Figure 13. Polar and dispersive components of the SFE and the W_A for the references and the coated samples.

3.5. 3D samples' Morphology

The SEM images of coated samples 2 and 7 (Table 1), acquired at different magnifications (125–500,000×) are displayed in Figures 14 and 15, respectively. It highlights that at the lowest magnification (125×), both samples exhibit characteristic sponge-like structures [15,19,32,34–36], with the formation of air pockets, which enhance the superhydrophobicity. Samples differ in the density of the sponge-like structures. In fact, it appears that sample 7, having higher roughness (i.e., 2.8 µm) generates air pockets that are more densely packed than the less rough sample 2 (Ra = 0.4 µm). In spite of this difference, both samples exhibit contact angles of 158° (see Figure 8). At the highest magnification (500,000×) (Figures 14 and 15), the hierarchical micro- and nano-structures appears, highlighting for both samples the dimension of the nanoparticles employed to formulate the coating, namely, about 30 nm in diameter. Therefore, in spite of the intrinsic roughness and the macroscopic density of the sponge-like structure, the generated air pockets made of multiscale roughness (micro- and nano-meter in dimension) guarantee the superhydrophobicity, i.e., contact angles of 158°.

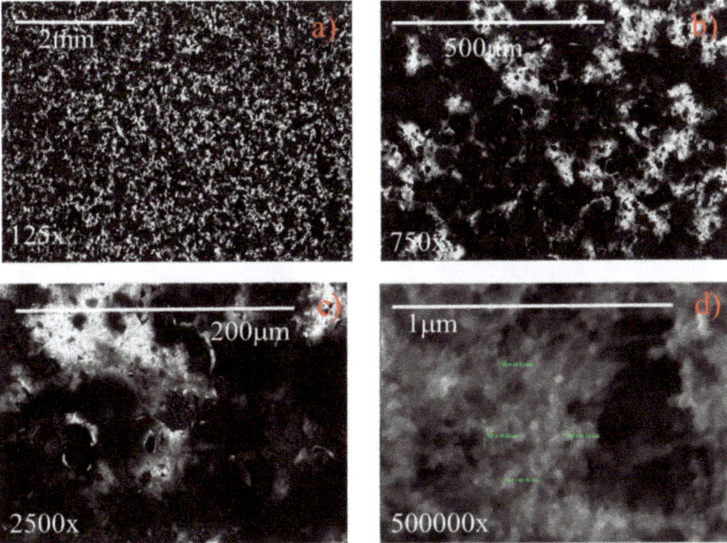

Figure 14. SEM images at different magnifications of a sample having an original roughness of sample 2 (**a**) 125×, (**b**) 750×, (**c**) 2500×, (**d**) 500,000×.

The surface 3D profilometry of coated samples 2 and 7, having actual roughness of 0.4 and 2.8 µm, respectively, were reported in Figure 16a,b; whereas Figure 16c,d show the corresponding references, having roughness of 0.8 and 4.0 µm, respectively. In the insert, the pictures representative of the water contact angles were also shown. It highlights that all samples exhibit a macroscopic needle-like morphology deriving from the specific textured roughness, and there are no differences in the needle-like morphology [36] for each series of samples (compare Figure 16a with Figure 16c, and Figure 16b with Figure 16d). The reference with low roughness (see Figure 16c) does not exhibit the Cassie state, the surface area in contact with the water droplet is high, and consequently the water contact angle is low (50°). When the reference roughness increases to about 4 µm (see Figure 16d), the relative distance between the surface irregularities and the related height allow the contact angle to increase to 115°. The increase of the water contact angle due to the increased roughness alone can be ascribed to the MORPHOLOGICAL EFFECT. By comparing coated samples with the corresponding references, it highlights that, although the needle-like morphology and the corresponding roughness do not change, the water contact angle increases from 50° and 115°, respectively, to 158°, as a consequence of the chemical modification of the surface (CHEMICAL EFFECT). Finally, by comparing the two coated samples (see Figure 16a,b), it must be noticed that, in spite of the different actual roughness, i.e., 0.4 and 2.8 μm, for the samples 2 and 7, respectively, both of them exhibit the same water contact angle, namely, 158°. Hence, in conclusion, it seems that the chemical effect prevails on the morphological effect.

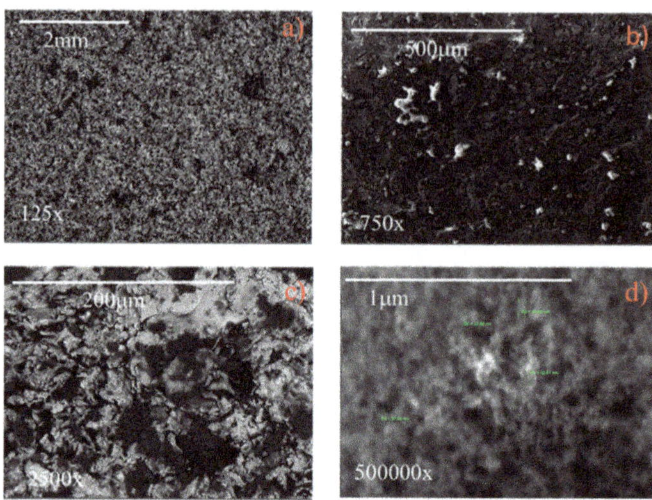

Figure 15. SEM images at different magnifications of a sample having an original roughness of sample 7 (**a**) 125×, (**b**) 750×, (**c**) 2500×, (**d**) 500,000×.

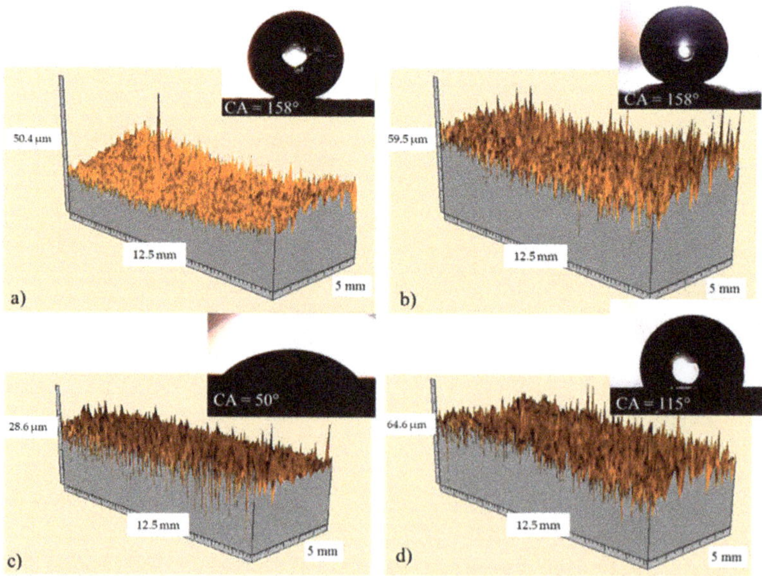

Figure 16. Surface 3D profilometry obtained with the laser for coated sample 2 (**a**) and sample 7 (**b**); in (**c**) and (**d**) the corresponding references were also reported. In the insert, the pictures representative of the related water contact angles are shown.

3.6. Cutting and Tape Test

Figure 17 shows images related to the cutting and tape test performed on coated sample 7. In detail, Figure 17a shows the picture after the cut and Figure 17b shows magnification of the surface after the test. The shape of the water droplet in Figure 17c highlights that the superhydrophobicity was preserved after the test too. It must be noted that, according to the ASTM D 3359-09 [30], the test was passed, since quite no detached flakes of coating could be observed. This result was classified as 5B [30].

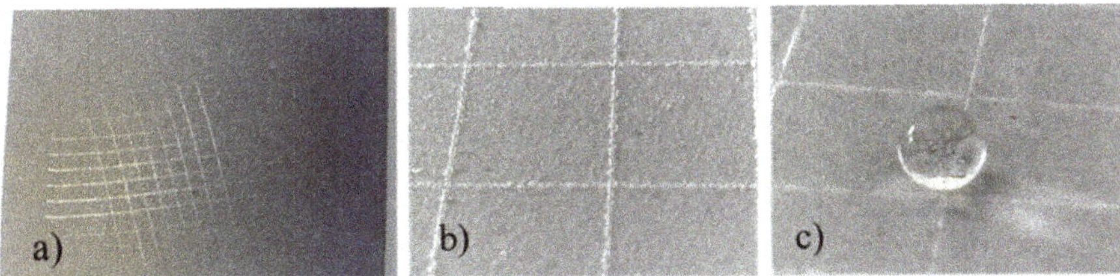

Figure 17. Cutting and tape test results on coated sample 7, (**a**) surface cut, (**b**) surface magnification, (**c**) surface wettability after test.

3.7. Low Temperature Wettability

A preliminary test about the wettability at low temperature of the developed coating was performed by freezing (at −27 °C) a few water droplets placed on both reference and coated sample 2 (pictures in Figure 18). It is interesting to note that the shape of the water droplets, and then of the wettability, does not change with freezing, and consequently, the coated samples display a reduced surface area in contact with the substrate if compared with the reference.

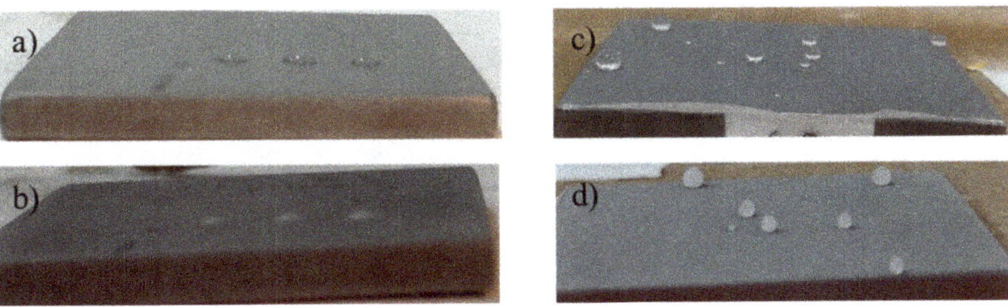

Figure 18. Pictures of samples acquired at −27 °C. Reference with water droplet before (**a**) and after (**b**) the icing at −27 °C; coated sample with water droplets before (**c**) and after (**d**) the icing at −27 °C.

4. Conclusions

In the present work, the authors studied the wettability of a new developed superhydrophobic coating applied with a layer-by-layer approach as a common aeronautical livery, that can be potentially employed as a passive anti-icing system for aeronautical applications.

It was found that the wettability of the uncoated samples decreases as the roughness increases achieving a value of about 115° for Ra = 4 μm. The increase of the water contact angle due to the increased roughness alone can be ascribed to the morphology changes (MORPHOLOGICAL EFFECT). Instead, high values of contact angle, i.e., about 158°, were achieved only after the application of the coating, regardless of the substrate roughness, ranging from 0.4 to 3 μm (CHEMICAL EFFECT), so highlighting that the chemical modifications prevail on the substrate morphology.

The coating's application reduced the SFE and W_A by 99% and 94%, respectively, with respect to the reference. Finally, a preliminary test about the wettability of the developed coating at low temperatures showed that the reduced wettability of coated samples was preserved also at −27 °C. In this regard, future research will be addressed to characterize the developed coating in severe environmental conditions simulating real flights.

Author Contributions: Conceptualization, F.P.; Data curation, F.P.; Formal analysis, F.P.; Funding acquisition, L.D.P.; Investigation, F.P., D.D. and F.M.; Methodology, F.P.; Project administration, A.C. and L.D.P.; Software, G.C.; Supervision, F.P. All authors have read and agreed to the published version of the manuscript.

Funding: This research was funded by H2020 Clean Sky 2 Framework (CS2), 807083—AIRFRAME ITD (GAM AIR 2018), Topic Identification Code: JTI-CS2-2015-CPW02-AIR-02-07, project number and acronym: 699757/SAT-AM (More Affordable Small Aircraft Manufacturing) and "The APC was funded by H2020 Clean Sky 2 Framework.

Acknowledgments: Giovanni Pucci (Sapienza University of Rome, Italy) is kindly acknowledged for his technical support.

Conflicts of Interest: The authors declare no conflict of interest.

References

1. Kreder, M.J.; Alvarenga, J.; Kim, P.; Aizenberg, J. Design of Anti-Icing Surfaces: Smooth, Textured or Slippery? *Nat. Rev. Mater.* **2016**, *1*, 15003–15039. [CrossRef]
2. Lv, J.; Song, Y.; Jiang, L.; Wang, J. Bio-inspired strategies for anti-icing. *ACS Nano* **2014**, *8*, 3152–3169. [CrossRef] [PubMed]
3. Schutzius, T.M.; Jung, S.; Maitra, T.; Eberle, P.; Antonini, C.; Stamatopoulos, C.; Poulikakos, D. Physics of Icing and Rational Design of Surfaces with Extraordinary Icephobicity. *Langmuir* **2015**, *31*, 4807–4821. [CrossRef] [PubMed]
4. Meuler, A.J.; Smith, J.D.; Varanasi, K.K.; Mabry, J.M.; McKinley, G.H.; Cohen, R.E. Relationship between water wettability and ice adhesion. *ACS Appl. Mater. Intefaces* **2010**, *2*, 3100–3110. [CrossRef] [PubMed]
5. Golovin, K.; Kobaku, S.; Lee, D.; Di Loreto, E.; Mabry, J.; Tuteja, A. Designing durable icephobic surfaces. *Sci. Adv.* **2016**, *2*, e1501496. [CrossRef]
6. Antonini, C.; Innocenti, M.; Horn, T.; Marengo, M.; Amirfazli, A. Understanding the effect of superhydrophobic coatings on energy reduction in anti-icing systems. *Cold Reg. Sci. Technol.* **2011**, *67*, 58–67. [CrossRef]
7. Alizadeh, A.; Yamada, M.; Li, R.; Shang, W.; Otta, S.; Zhong, S.; Ge, L.; Dhinojwala, A.; Conway, K.R.; Bahadur, V.; et al. Dynamics of Ice Nucleation on Water Repellent Surfaces. *Langmuir* **2012**, *28*, 3180–3186. [CrossRef]
8. Simpson, J.T.; Hunter, S.R.; Aytug, T. Superhydrophobic materials and coatings: A review. *Rep. Prog. Phys.* **2015**, *78*, 086501–086515. [CrossRef]
9. Lin, Y.; Chen, H.; Wang, G.; Liu, A. Recent Progress in Preparation and Anti-Icing Applications of Superhydrophobic Coatings. *Coatings* **2018**, *8*, 208. [CrossRef]
10. Fresnais, J.; Chapel, J.P.; Benyahia, L.; Poncin-Epaillard, F. Plasma-treated superhydrophobic polyethylene surfaces: Fabrication, wetting and dewetting properties. *J. Adhes. Sci. Technol.* **2009**, *23*, 447–467. [CrossRef]
11. Hikita, M.; Tanaka, K.; Nakamura, T.; Kajiyama, T.; Takahara, A. Super-liquid repellent surfaces prepared by colloidal silica nanoparticles covered with fluoroalkyl groups. *Langmuir* **2005**, *21*, 7299–7302. [CrossRef] [PubMed]
12. Yang, H.T.; Jiang, P. Self-cleaning diffractive macroporous films by doctor blade coating. *Langmuir* **2010**, *26*, 12598–12604. [CrossRef] [PubMed]
13. Bravo, J.; Zhai, L.; Wu, Z.Z.; Cohen, R.E.; Rubner, M.F. Transparent superhydrophobic films based on silica nanoparticles. *Langmuir* **2007**, *23*, 7293–7298. [CrossRef] [PubMed]
14. Asadollahi, S.; Farzaneh, M.; Stafford, L. On the Icephobic Behavior of Organosilicon-Based Surface Structures Developed Through Atmospheric Pressure Plasma Deposition in Nitrogen Plasma. *Coatings* **2019**, *9*, 679. [CrossRef]
15. Liu, J.; Janjua, Z.A.; Roe, M.; Xu, F.; Turnbull, B.; Choi, K.S.; Hou, X. Super-Hydrophobic/Icephobic Coatings Based on Silica Nanoparticles Modified by Self-Assembled Monolayers. *Nanomaterials* **2016**, *6*, 232. [CrossRef]
16. Cao, L.; Jones, A.K.; Sikka, V.K.; Wu, J.; Gao, D. Anti-icing superhydrophobic coatings. *Langmuir* **2009**, *25*, 12444–12448. [CrossRef]

17. Kustas, F.M.; Kustas, A.B.; Williams, T.S.; Hicks, R. Fabrication of Superhydrophobic and Icephobic Coatings by Nanolayered Coating Method. U.S. Patent Application 20190127841 A1, 19 September 2019.

18. Furstner, R.; Barthlott, W.; Neinhuis, C.; Walzel, P. Wetting and self-cleaning properties of artificial superhydrophobic surfaces. *Langmuir* **2005**, *21*, 956–961. [CrossRef]

19. Piscitelli, F.; Tescione, F.; Mazzola, L.; Bruno, G.; Lavorgna, M. On a simplified method to produce hydrophobic coatings for aeronautical applications. *Appl. Surf. Sci.* **2019**, *472*, 71–81. [CrossRef]

20. *ISO 4288—Geometrical Product Specifications (GPS)–Surface Texture: Profile Method–Rules and Procedures for the Assessment of Surface Texture*; International Organization for Standardization: Geneva, Switzerland, 1996.

21. Young, T. An Essay on the Cohesion of Fluids. *Philos. Trans. R. Soc. Lond.* **1805**, *95*, 65–87. [CrossRef]

22. D7490-13 Standard Test Method for Measurement of the Surface Tension of Solid Coatings, Substrates and Pigments using Contact Angle Measurements. American Society for Testing and Materials: Conshohocken, PA, USA, 7 January 2013.

23. Owens, D.K.; Wendt, R.C. Estimation of the surface free energy of polymers. *Appl. Surf. Sci.* **1969**, *13*, 1741–1747. [CrossRef]

24. Żenkiewicz, M. Methods for the calculation of surface free energy of solids. *J. Achiev. Mater. Manuf. Eng.* **2007**, *24*, 137–145.

25. Rudawska, A.; Jacniacka, E. Analysis of Determining SFE Uncertainty with the Owens-Wendt method. *Int. J. Adhes. Adhes.* **2009**, *29*, 451–457. [CrossRef]

26. Models for Surface Free Energy Calculation. Available online: https://www.kruss.de/fileadmin/user_upload/website/literature/kruss-tn306-en.pdf (accessed on 26 December 2019).

27. Wenzel, R.N. Resistance of Solid Surfaces to Wetting by Water. *Ind. Eng. Chem. Res.* **1936**, *28*, 988–994. [CrossRef]

28. Banerjee, S. Simple derivation of Young, Wenzel and Cassie–Baxter equations and its interpretations. *arXiv* **2008**, arXiv:0808.1460.

29. Cassie, A.B.D.; Baxter, S. Wettability of porous surfaces. *Trans. Faraday Soc.* **1944**, *40*, 546–551. [CrossRef]

30. *ASTM D3359–09 Standard Test Methods for Measuring Adhesion by Tape Test*; ASTM: West Conshohocken, PA, USA, 2009.

31. Kam, D.H.; Bhattacharya, S.; Mazumder, J. Control of the wetting properties of an AISI 316L stainless steel surface by femtosecond laser induced surface modification. *J. Micromech. Microeng.* **2012**, *22*, 105019. [CrossRef]

32. Yang, C.; Tartaglino, U.; Persson, B.N.J. Influence of Surface Roughness on Superhydrophobicity. *Phys. Rev. Lett.* **2006**, *97*, 116103. [CrossRef]

33. Zheng, Q.; Lu, C. Size effect of surface roughness to Superhydrophobicity. *Procedia IUTAM* **2014**, *10*, 462–475. [CrossRef]

34. Lai, Y.K.; Chen, Z.; Lin, C.J. Recent Progress on the Superhydrophobic Surfaces with Special Adhesion: From Natural to Biomimetic to Functional. *J. Nanoeng. Nanomanuf.* **2011**, *1*, 1–17. [CrossRef]

35. Amigoni, S.; de Givenchy, E.T.; Dufay, M.; Guittard, F. Covalent Layer-by-Layer Assembled Superhydrophobic Organic–Inorganic Hybrid Films. *Langmuir* **2009**, *25*, 11073–11077. [CrossRef]

36. Farhadi, S.; Farzaneh, M.; Kulinich, S.A. Anti-icing performance of superhydrophobic surfaces. *Appl. Surf. Sci.* **2011**, *257*, 6264–6269. [CrossRef]

Article

Assessment of Aircraft Surface Heat Exchanger Potential

Hagen Kellermann *, Anaïs Luisa Habermann and Mirko Hornung

Bauhaus Luftfahrt e.V., Willy-Messerschmitt Straße 1, 82024 Taufkirchen, Germany
* Correspondence: hagen.kellermann@bauhaus-luftfahrt.net

Received: 20 November 2019; Accepted: 12 December 2019; Published: 19 December 2019

Abstract: Providing sufficient cooling power for an aircraft will become increasingly challenging with the introduction of (hybrid-) electric propulsion. To avoid excessive drag from heat exchangers, the heat sink potential of the aircraft surface is evaluated in this study. Semi-empirical correlations are used to estimate aircraft surface area and heat transfer. The impact of surface heating on aircraft drag is qualitatively assessed. Locating surface heat exchangers where fully turbulent flow is present promises a decrease in aircraft drag. Surface cooling potential is investigated over a range from small regional aircraft to large wide body jets and a range of surface temperatures. Four mission points are considered: Take-off, hot day take-off, climb and cruise. The results show that surface heat exchangers can provide cooling power in the same order of magnitude as the waste heat expected from (hybrid-) electric drive trains for all sizes of considered aircraft. Also, a clear trend favouring smaller aircraft with regards to the ratio of available to required cooling power is visible.

Keywords: aircraft thermal management; hybrid electric propulsion; surface heat exchanger

---

1. Introduction

Research for next generation commercial aircraft is driven by ambitious goals to reduce the aircraft's environmental impact such as the European Commission's Strategic Research and Innovation Agenda (SRIA) [1] that targets a 75% reduction in CO_2 emissions by the year 2050 compared to the year 2000. A big contributor to achieve those targets is the propulsion system. Novel propulsion concepts with intercoolers [2], topping cycles [3] or bottoming cycles [4] are currently under investigation to reduce the specific fuel consumption. Another promising approach seems to be a higher electrification of the on-board systems or even the propulsion system. Examples are the more electric aircraft [5] or (hybrid-) electric propulsion systems [6]. Their electric components generate waste heat that needs to be rejected in an efficient way. Many concepts result in higher thermal loads of the systems. Conventional cooling concepts require ram air and heat exchangers, which are placed in the airflow path and thus generate drag [7]. Another option is to use existing aircraft surfaces for heat transfer from the inside of the aircraft to the ambient [8]. These structurally integrated heat exchangers may be beneficial for both weight and drag of the Thermal Management System (TMS) because no additional components such as the ram air heat exchanger are required and no components are installed in the flow path. Additionally, heat rejection to the aircraft's boundary layer may lead to drag reductions [9]. The aim of this paper is to investigate the heat sink potential of available aircraft surfaces.

Wang et al. present a good overview of the application of surface heat exchangers in aircraft up to the year 1999. In the beginning, the development of surface heat exchangers was driven by the cooling demand of piston engines with increasing power densities. In the 1920ies and 1930ies they were mainly used in racing aircraft. In some aircraft such as the "Supermarine S.6" surface heat exchangers covered surfaces of multiple components such as wings, fuselage and floats. In military aircraft, leading edge steam radiators were successfully tested. However, despite the proven thermodynamic performance the technology was not put into practical applications due to hazards such as machine gun fire. When gas turbines started to replace piston engines in aircraft, the engine cooling problem vanished and with it surface heat exchangers. However, academic research on surface heating continued. The results indicate that heating aircraft surfaces might not only serve for heat dissipation but also as means of boundary layer control. It is commonly agreed that heat addition to a laminar boundary layer increases instabilities and therefore may lead to an earlier transition, thus increasing drag [8].

More recent studies showed a growing interest in surface heat exchangers again due to the increased cooling demand from the aforementioned technologies. Especially new engine concepts such as Ultra High Bypass Ratio Turbofans and open rotors with very compact gas generators and mechanical transmission have increased oil heat loads. Sousa et al. investigated a surface cooler with fins inside a turbo fan engine bypass as air cooled oil cooler (ACOC). Numerical calculations in combination with experiments were conducted. They showed that the surface cooler was capable of rejecting 76% of the take-off oil heat load [10]. Surface air cooled oil coolers (SACOC) are investigated by multiple EU-funded projects such as SHEFAE [11], SHEFAE 2 [12] and SACOC [13]. Sakuma et al. carried out investigations of the effects of varying SACOC geometries in the context of SHEFAE 2. They found that two 200 mm long heat exchangers could reject the same heat as one 900 mm long one while maintaining the same pressure drop allowing for area and weight optimization of the SACOC [14].

Recently, Liu et al. conducted numerical studies to describe pressure loss and heat transfer of different aircraft surface heat exchanger fin configurations including continuous, segmented and staggered fins. They found the continuous configuration to have the most advantageous heat transfer to pressure drop ratio [15]. Part of the wing surfaces were used for heat dissipation of a hybrid electric aircraft with a TMS utilizing fuel as working fluid. The results showed that steady state cooling of the electric propulsion system is possible in most operating points, however the aircraft only had 20% hybridization [16].

While there is a good amount of literature on surface coolers in aircraft applications, most investigate research questions tailored to one specific engine or aircraft or try to optimize the surface heat exchanger geometry. In contrast, this study aims to generally predict the thermodynamic potential of the aircraft surface for a range of differently sized aircraft. The goal is to quickly assess the feasibility of using a TMS with surface heat exchangers for a hybrid electric configuration.

For that purpose, a thermodynamic model of a surface heat exchanger covering existing aircraft surfaces is developed. A scalable geometric model of a tube and wing type aircraft with podded engines is derived with a semi-empirical approach. It is used to analyse the impact of aircraft size and available portions of the total surface area on the potential cooling power. Various sensitivies of the model including surface temperature, incoming radiation and component geometries are considered. In addition, drag increments resulting from non-adiabatic boundary layers are assessed.

The ambient conditions differ at each operating point. The study evaluates steady state heat transfer performance in pre-defined sets of Mach number (Ma), altitude (alt) and ISA temperature deviation (dT_{ISA}). They reflect typical operating conditions of commercial aircraft namely Take-Off (TO), Hot Day Take-Off (HTO), Climb (CL) and Cruise (CR), which are relevant sizing points for the TMS. The quantification of the potential of the aircraft's skin as heat sink can be used by future projects on advanced propulsion concepts to account for the total amount or a fraction of the system's waste heat removal.

2. Aircraft Correlations

The study aims to estimate the surface heat sink potential of a range of aircraft covering most of the commercial aviation market. Therefore, data for aircraft ranging from small regional aircraft up to large wide body jets are used as basis for the correlations. Besides correlations for the wetted surface area (A_{wet}), the data is analysed with regard to propulsive power as it will be an indicator for the size of future hybrid electric power trains and thus the expected required cooling power (Q_{rq}). Most data are obtained from Reference [17]. They provide aircraft data up to the year 2000 for aircraft from different manufactures including Airbus, Boeing, Fokker and Bombardier. Additional data especially for newer aircraft are extracted from documents provided by the manufacturers [18,19].

2.1. Aircraft Component Geometries

A_{wet} of the aircraft consists of the surface areas of multiple components. This study is strictly limited to the tube and wing aircraft configuration and thus the components considered as possible locations for surface heat exchangers are:

- Fuselage
- Wing
- Nacelles
- Horizontal tail
- Vertical tail

For wing, horizontal and vertical tail the data at hand contains the exposed area (A_{exp}) that is, for a wing the area given is the base area outside the fuselage. In a first order approximation, A_{exp} is doubled to calculate A_{wet}. More accurate semi-empirical methods to calculate the wetted area of bodies with an airfoil cross section for example in Reference [20] exist, however, for an initial potential assessment, it seems more reasonable to choose the simplest method possible. For the fuselage and nacelles, A_{wet} is also not directly available. Instead, length and diameter (in case of the nacelle the maximum diameter) are included. A_{wet} of these components is estimated by using the geometric model of a cylinder. This approach overestimates the area for the nacelles, because a cylinder with the nacelle length and the maximum diameter as constant diameter has a larger A_{wet} than the actual nacelle with a variable diameter. For the fuselage, the overestimation of the lateral surface area is reduced by the fact that an open cylinder model is used but the fuselage is actually a closed body. For a quick estimation of the order of magnitude of the error from these geometric simplifications, a point validation is conducted using available data from an A320 sized aircraft model [21]. Table 1 shows the comparison between the simplified A_{wet} (A_{sim}) and the actual A_{wet} (A_{act}) as well as the relative deviation of A_{sim} from A_{act}:

Table 1. Comparison of simplified A_{wet} with actual A_{wet}.

Component	$A_{sim}\,(\mathrm{m}^2)$	$A_{act}\,(\mathrm{m}^2)$	$\Delta_{A,wet}\,(\%)$
Fuselage	478.0	412.9	+15.8
Wing	202.4	208.7	−3.0
Nacelles	55.8	52.4	+6.4
Horizontal tail	48.4	49.6	−2.4
Vertical tail	42.2	43.3	−2.7
Total	826.8	766.9	7.8

The deviation for the total A_{wet} is less than 10%. For each component, the expected direction of deviation is confirmed that is, for fuselage and nacelles the simplifications lead to an overestimation of A_{wet} whereas for all the other components the methods underestimate A_{wet}. The largest deviation is present for the fuselage with 15.8%. Overall, the deviations are considered acceptable for the scope of this study, because the aim is to find basic correlations among a wide range of aircraft rather than developing precise calculation methods for one specific aircraft.

2.2. Surface Area Correlations

Four possible aircraft design parameters are identified as potential variables to correlate with the total wetted surface area:

1. Maximum Take-Off Weight ($MTOW$)
2. Maximum number of seats (n_{max})
3. Maximum payload (MPL)
4. Design range (R_{des})

All four are defined in the conceptual design stage of an aircraft and affect the overall aircraft design. For all four variables, correlations with A_{wet} are found using least squares polynomial fits. To assess the quality of each fit, the coefficient of determination (r^2) is used. The best fits that is, the ones with the highest r^2 value for all four variables are linear fits of the log-log scaled data and are shown in Figure 1. The corresponding fits are summarized in Equation (1) with the coefficients given in Table 2.

$$log_{10}A_{wet} = a \cdot log_{10}x + c \tag{1}$$

Table 2. Coefficients for log-log surface area fits.

x	a	c	r^2
$MTOW$	0.748	−0.689	0.986
n_{max}	0.940	0.887	0.963
MPL	0.855	−0.668	0.965
R_{des}	0.995	−0.417	0.859

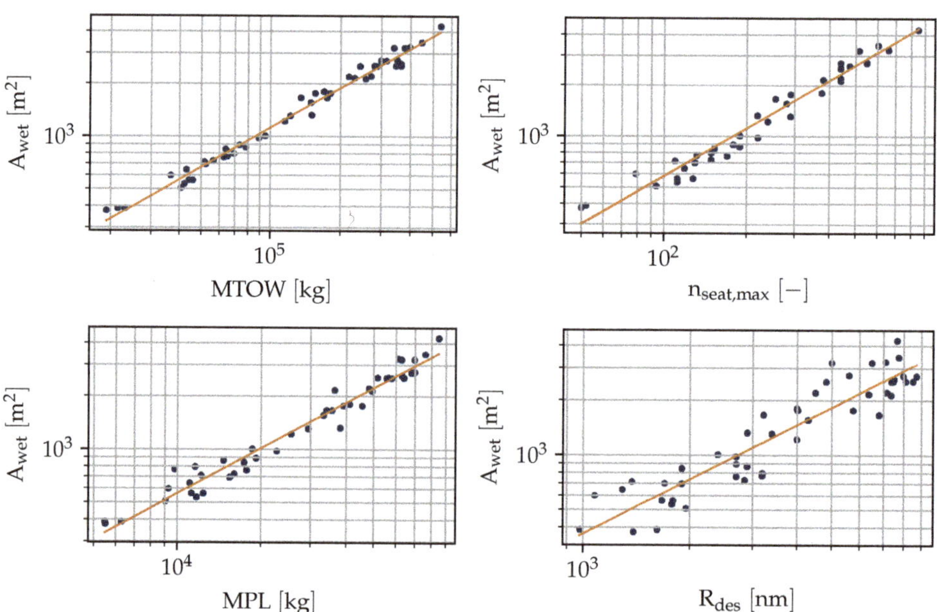

Figure 1. A_{wet} correlations with data from References [17–19].

The correlation of A_{wet} with R_{des} has a higher variance than the other three. A_{wet} correlates very well with $MTOW$, MPL and n_{max} ($r^2 > 0.95$). For this study, the $MTOW$ correlation is chosen because it has the highest r^2 value and $MTOW$ is the most general and robust aircraft parameter to compare against. It could also be used for retro fitted cargo aircraft, which is not possible for n_{max}. However, the other correlations might be useful for a first A_{wet} assessment prior to the *MTOW* calculation in the conceptual design phase. The correlations are limited to their source data that is, they may not be used outside the range of the source data. They may be used for future aircraft that is, hybrid electric aircraft if no significant change in the respective correlation is expected due to for example technology changes. From the heat transfer modelling (cf. Section 3) it becomes apparent that solely knowing the total wetted area of the aircraft ($A_{wet,tot}$) is not sufficient even for the simple correlations that are used in this study. The distribution of $A_{wet,tot}$ among the component groups mentioned in Section 2.1 is investigated. No correlation is found with any of the x-parameters from Table 2. The share of each component of $A_{wet,tot}$ ($A_{wet,i}$) is rather constant. Therefore, a mean is applied and the results are listed together with the standard deviation (σ) for each mean in Table 3.

Table 3. $A_{wet,i}\,/\,A_{wet,tot}$ for each component.

Component	$A_{wet,i}/A_{wet,tot}$ (%)	σ (%)
Fuselage	49	3.80
Wing	31	3.55
Nacelles	7	1.49
Horizontal Tail	8	1.21
Vertical Tail	5	0.83

2.3. Propulsive Power

To put the available cooling power (Q_{av}) in perspective with the required cooling power (Q_{rq}) an estimation of the expected waste heat is necessary. The quantity of waste heat of a future propulsion system will depend on many factors, especially the propulsive power (P_{prop}), the transmission efficiency (η_{trans}), which includes all losses from shaft power (P_{shaft}) to P_{prop}, the Degree of Power Hybridization (H_P) [22] of the drive train and the overall electric efficiency (η_{ec}). Calculation of the exact heat loads over the entire mission are part of a detailed iterative design process. For a first estimation, simple methods are applied to estimate the waste heat during take-off, which is likely to be one of the most critical mission points with regards to cooling requirements. Starting from the take-off thrust (F_{TO}), which is available in the data set, Q_{rq} is derived:

$$Q_{rq} = (1 - \eta_{trans}) \cdot P_{prop} \cdot H_P \cdot (1 - \eta_{ec}) \tag{2}$$

with P_{prop} as:

$$P_{prop} = F_{TO} \cdot v_{TO} \tag{3}$$

With v_{TO} being the take-off velocity, which is calculated based on Sea Level (SL) conditions with $dT_{ISA} = 0$ and $Ma = 0.2$ as representative Ma_{TO}. The F_{TO} values are obtained from another linear fit of the log-log scaled data over *MTOW*. The resulting fit (Equation (4)) is shown in Figure 2. It has an r^2 value of 0.983.

$$log_{10}F_{TO} = 0.913 \cdot log_{10}MTOW + 0.895 \tag{4}$$

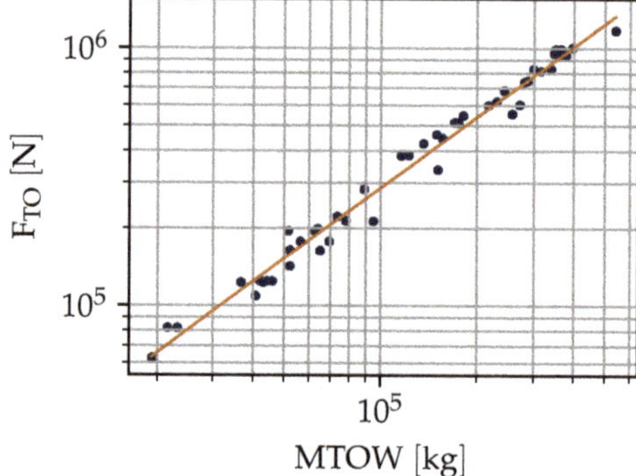

Figure 2. MTOW–F_{TO} correlation with data from References [17–19].

3. Surface Heat Transfer

In this section, the applied heat transfer models are described. The sensitivity of the methods to changes in geometry is tested and the impact of surface heating on drag is assessed.

3.1. Modeling

Flat plate models with uniform temperature distribution are used for heat transfer calculations for all components. Correlations for the local Nusselt number (Nu_x) from References [23–25] are applied to calculate local heat transfer coefficients (α_x), which requires a local discretization of the geometry in flow direction. The trapezoidal shaped components (wing and tail planes) are also discretized in span wise direction to account for the different flow lengths and corresponding Reynolds numbers (Re_x). Incoming solar radiation is accounted for in the overall heat balance by means of a material absorption coefficient and an incoming radiation power (P_{rad}) on all surfaces that are exposed to the sun. Unless stated otherwise an absorption coefficient of 0.25 typical of white paint and P_{rad} of 1362 W/m², which is the constant value outside earth's atmosphere [26] are assumed. For each component, half of A_{wet} is considered to be exposed to P_{rad}. Those are conservative assumptions since P_{rad} has a slightly lower value even at the highest flight levels than the above-mentioned value outside the atmosphere. Detailed descriptions of the used convection correlations and heat balances can be found in Reference [16]. The used 2D methods are less precise than for example 3D Computational Fluid Dynamics methods but they are sufficient for a first quantification of the surface cooling power in the conceptual design stage.

3.2. Sensitivities

The aforementioned local discretization of the heat transfer calculation depends on Re_x. Section 2.2 focuses on correlations of the total and component wise A_{wet}. However, to calculate Re_x more knowledge of the geometry is required. For example, two fuselages with the same A_{wet} have different Re_x distributions if their slenderness ratios (Λ) differ. To account for these effects, the geometric model of the components is refined. For cylindrical components (fuselage, nacelle) the sensitivity of Λ is studied:

$$\Lambda = l/d \tag{5}$$

With length (l) and diameter (d). Wing and tail components are modelled as single section trapezoids with no leading edge sweep. Their geometries, specifically the span-wise chord distribution can be fully defined with the help of their A_{exp}, aspect ratio (AR) and taper ratio (λ) [20]. The following sensitivity studies are conducted around TO conditions. It is one of the most critical conditions for TMS design, because of the low air flow velocities, high ambient temperatures (T_{amb}) and large cooling demand (Q_{rq}) due to maximum propulsive power. Unless otherwise specified, the values in Table 4 are assumed for the wing sensitivity studies. The values are not specific to any aircraft but generally lie inside the range of the given data. $Re_{x,c}$ is the critical Reynolds number and T_{surf} the average surface temperature. In a real cooling application, the surface temperature would most likely not be uniform but have a gradient in the direction of a hot side flow underneath the surface. However, in this first approximation an average T_{surf} is assumed for simplification.

Table 4. Wing sensitivity study parameters.

Parameter	Value
A_{exp}	200 m^2
AR	12
λ	0.29
$Re_{x,c}$	5×10^5
T_{surf}	320 K

3.2.1. Transition Location

Flat plate heat transfer correlations distinguish between laminar and turbulent flow. They rely on the knowledge of a critical location (x_c) where transition occurs. Usually x_c is defined by $Re_{x,c}$ which according to Reference [24] is between 1×10^5 to 3×10^6 depending on free stream turbulence and surface roughness. Detailed transition modelling is a complex research area and beyond the scope of this work. However, a sensitivity study with varying $Re_{x,c}$ and Ma ranging from slow taxiing $Ma = 0.01$ to representative TO $Ma = 0.2$ is conducted. The results of the theoretically available cooling power (Q_{av}) as well as the relative Q_{av} compared to the Q_{av} at the lowest $Re_{x,c}$ for each Ma ($Q_{Re,x,c,min}$) are displayed in Figure 3.

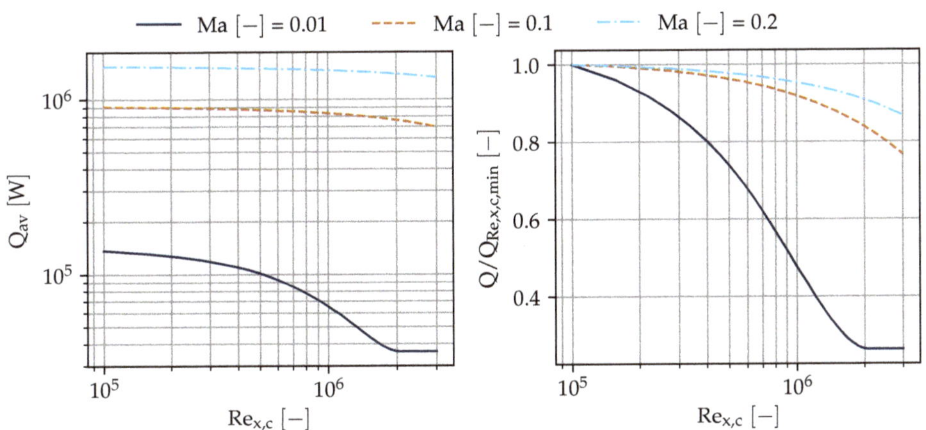

Figure 3. Transitional Reynolds number sensitivity.

An increased Ma results in increased Q_{av} because of the increased effects of forced convection. Shifting $Re_{x,c}$ to higher values, results in a decrease in Q_{av}. Turbulent flows favour heat transfer more than the structured flow in laminar regions because of the increased particle mixing within the boundary layer. With increased $Re_{x,c}$ the portion of A_{exp} with laminar flow increases. For very low Ma increasing $Re_{x,c}$ beyond 2×10^6 results in laminar flow on the entire surface. A further increase in $Re_{x,c}$ has no additional effect. The transition point has a large influence on Q_{av}. For 3D wings, transition is more complex than defining an $Re_{x,c}$ and assuming instantaneous transition. This study does not accurately account for real transition effects. The results in Section 4 assume fully turbulent flow areas downstream the transition location. Therefore, the results of this study cannot directly be used for concepts with enhanced laminarity such as natural laminar flow (NLF) wings. Covering these advanced aerodynamic concepts is part of future work.

3.2.2. Wing Aspect Ratio

AR is varied from 6 to 18—a range that includes all aircraft used for the correlations in Section 2 and also leaves margin for possible future aircraft with increased AR. The results of Q_{av} as well as the relative Q_{av} compared to the Q_{av} at the lowest AR for each Ma ($Q_{AR,min}$) are displayed in Figure 4. For better comprehension of the trends in Figure 4, the effect of increasing AR on the local α_x distribution for the lowest and largest Ma are illustrated with heat maps in Figure 5.

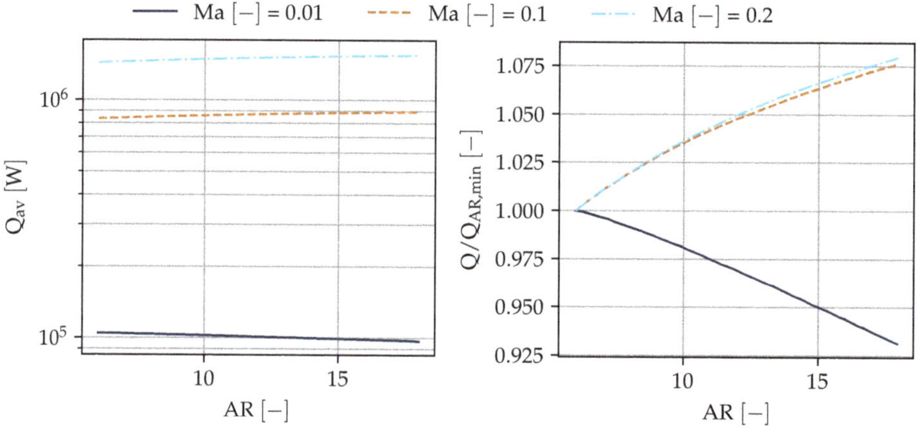

Figure 4. Wing aspect ratio sensitivity.

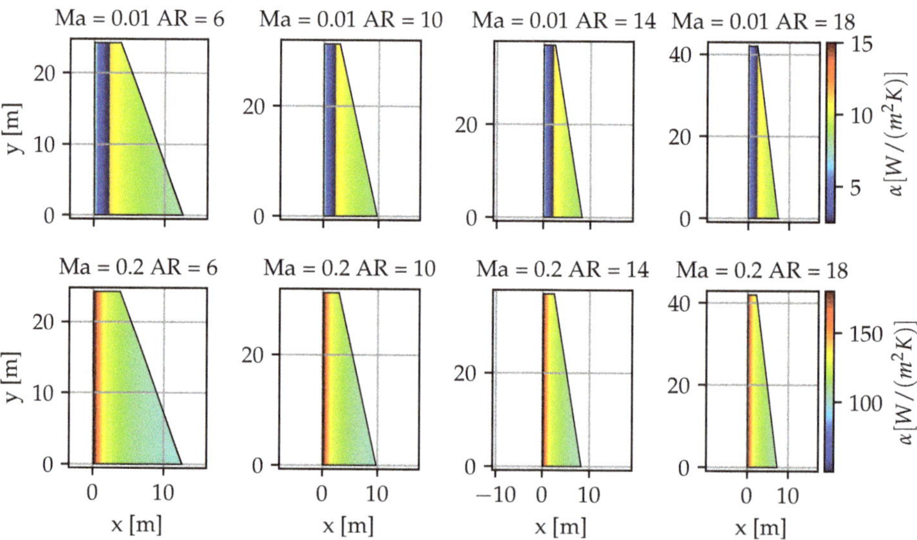

Figure 5. Local heat transfer coefficient for different Mach number–wing aspect ratio combinations.

Q_{av} increases with Ma, because the forced convection increases, due to increasing Re. Depending on Ma, Q_{av} increases or decreases with increasing AR. More specifically for the lowest Ma of 0.01, Q_{av} decreases with increasing AR. For all other Ma used in the study Q_{av} increases with AR. Two counteracting effects are the reason:

1. In general, α_x decreases along x because of the increasing thickness of the thermal boundary layer (δ_T). Therefore, higher AR favours heat transfer because for the same area, the average chord length is lower (cf. Figure 5 bottom graphs).
2. The front section of the wing is laminar, which results in small α_x. A higher AR increases the span and, thus, the laminar portion of the plate's total area (cf. Figure 5 top two graphs). The x_c depends on Ma. For low Ma the transition occurs further downstream, which means that this second effect contributes more.

Tripling the aspect ratio results in $\pm 8\%$ Q_{av} depending on Ma. The sensitivity is too weak for the expected precision of this study that aims to determine the order of magnitude of the surface cooling power. Hence, it is not regarded in the following studies.

3.2.3. Wing Taper Ratio

λ is varied between 0.1 and 1.0. The same Ma range as in the previous sections is applied. Variations in Q_{av} do not exceed $\pm 2\%$ with slight advantages for the non-tapered wings ($\lambda = 1.0$). The aforementioned effect of increasing flow length is positive for heat transfer of tapered wings near the wing tip but negative near the root, which leads to its equalization after integration over the entire span. As with the AR sensitivity, the effect is too small to be further considered in this work.

3.2.4. Fuselage Slenderness Ratio

For a fuselage with $A_{wet} = 1600$ m^2, Λ is varied from 5 to 15 within the same Ma range as the previous sensitivity analysis. Regardless of Ma, the change in Q_{av} from the lowest to the highest Λ value is around -8%, again due to the increasing flow length with increasing Λ. The effect is also within the expected precision of this study. For further investigations, $\Lambda = 12$ is used, which is conservative as it is one of the highest Λ values found in today's aircraft for example, for the Airbus A340-600.

3.3. Drag

For any aircraft component, which contributes to the aircraft's drag, local surface temperature can influence the aerodynamics of air passing the component surface at a certain velocity and with certain fluid characteristics. The two main occurring effects depending on the fluid's initial state are:

1. Transition delay of initially laminar flow
2. Drag alteration of fully turbulent flow

As the skin friction coefficient (c_f) is significantly smaller in laminar than in turbulent flow, total skin friction drag (D_f) of a surface can be decreased by moving the transition location downstream that is, by increasing the laminar length. During the last centuries, laminar flow control approaches have been studied intensively as a means to decrease drag. As such, surface temperature alteration can be employed to decrease the growth rate of unstable disturbances in the fluid and thus, to repress transition from laminar to turbulent flow [27]. The application of this method was shown in experiments by for example, References [27,28]. Two different approaches apply [29]:

1. Heating/cooling of the whole wetted surface area
2. Strategic heating/cooling of a part of the wetted surface area

In the two-dimensional case, Tollmien-Schlichting instabilities, which dominate the laminar boundary layer, are mitigated by cooling of the near wall boundary layer. In accordance with theory, flat plate experiments showed that the cooling of a surface leads to an increase of $Re_{x,c}$ and a downstream movement of x_c [28]. The effect is reversed when the surface is heated: the destabilizing effect of the temperature increase in the boundary layer dominates and x_c moves upstream [27].

However, the stabilizing effect of cooling can also be utilized when a portion of the surface is heated at strategic locations. For a two-dimensional case, it was shown that heating a portion of a surface where stable laminar flow is present (preferably the leading edge) followed by a cool that is, unheated, "relaxation" surface downstream can lead to a preferable downstream movement of x_c. The heated wall has to be situated in the region where Tollmien-Schlichting waves start to develop in the laminar boundary layer. The temperature of the near wall boundary layer is increased and when the fluid reaches the cooler wall further downstream, the temperature of the boundary layer is higher than the wall temperature. The boundary layer is cooled down and the growth rate of the unstable disturbances is decreased. The transition point moves downstream. If the surface is heated in an unstable flow region, the effect is reversed [27,29,30].

In three-dimensional airflows, however, cross-flow instabilities determine the boundary layer. Dovgal et al. showed that in this case, a temperature increase of the near wall boundary layer fosters cross-flow instabilities no matter if the whole surface or only a part of the surface is heated. The transition location moves upstream resulting in an increased D_f [27,30]. Thus, for any three-dimensional aircraft component, localized and global surface heating in the laminar flow region facilitates laminar to turbulent transition and increases D_f.

In contrast, when the boundary layer is fully turbulent, different mechanisms govern the flow: Heating of the near wall boundary layer reduces the turbulent D_f. Kramer et al. conducted wind tunnel experiments and flight tests in 1999. They found that an increase of the near wall boundary layer temperature leads to a decrease of Re_x, which in turn leads to a reduced local skin friction force [31]. For a body similar to a fuselage, they showed that the heating of the fore body leads to a higher drag reduction than the heating of the aft body, whereas the heating of the whole body has the highest drag reduction potential. The findings are supported by a numerical evaluation of the effect of heating on the turbulent boundary layer flow over slender and bluff fuselage-like bodies conducted by Lin and Ash in 1986 [32]. The following theoretical deviation of D_f as a function of wall heating for a smooth flat plate is based on the deviation proposed by Reference [31].

The length Reynolds number is defined as:

$$Re_x = \frac{\rho \cdot v \cdot x}{\mu} \tag{6}$$

For $Re_x = 10^6 - 10^8$, the turbulent c_f for a flat plate of length x can be expressed with $K = 0.036$, $m = 6$ by Reference [33]:

$$c_f = \frac{K}{Re_x^{\frac{1}{m}}} \tag{7}$$

Total skin friction drag of a flat plate with the length x and total area A for a turbulent boundary layer is defined as [33]:

$$D_f = c_f \cdot \frac{1}{2} \cdot \rho \cdot v^2 \cdot A = \frac{0.036}{2} \cdot \frac{\rho}{\rho^{\frac{1}{6}}} \cdot \frac{v^2}{v^{\frac{1}{6}}} \cdot x^{\frac{1}{6}} \cdot \mu^{\frac{1}{6}} \cdot A. \tag{8}$$

Assuming a constant heated surface temperature (T_h) along x, the temperature ratio of unheated air (T_u) and T_h is defined as:

$$TR = \frac{T_h}{T_u} \tag{9}$$

Applying the ideal gas law leads to $\rho = f(1/T)$ and the dynamic viscosity of air can be simplified to $\mu = f(T)$. Thus:

$$\frac{\rho_h}{\rho_u} = \frac{T_u}{T_h} = \frac{1}{TR} \tag{10}$$

$$\frac{\mu_h}{\mu_u} = \frac{T_h}{T_u} = TR \tag{11}$$

$$\frac{Re_{x,h}}{Re_{x,u}} = \frac{\rho_h \cdot v_h \cdot \mu_u}{\rho_u \cdot v_u \cdot \mu_h} = \left(\frac{1}{TR}\right)^2 \tag{12}$$

When the wall is heated ($TR > 1$), Re_x decreases with increased temperature. In consequence, c_f increases. However, the change in ρ has a larger effect on D_f than the change in c_f:

$$\frac{D_{f,h}}{D_{f,u}} = \frac{\rho_h^{\frac{5}{6}}}{\rho_u^{\frac{5}{6}}} \cdot \frac{\mu_u^{\frac{1}{6}}}{\mu_h^{\frac{1}{6}}} = \left(\frac{1}{TR}\right)^{\frac{5}{6}} \cdot TR^{\frac{1}{6}} = \left(\frac{1}{TR}\right)^{\frac{2}{3}} \tag{13}$$

and therefore if $T_h > T_u \rightarrow D_{f,h} < D_{f,u}$. The higher the wall temperature compared to the ambient temperature, the higher the drag decreasing potential. All simplified relations are depicted in Figure 6.

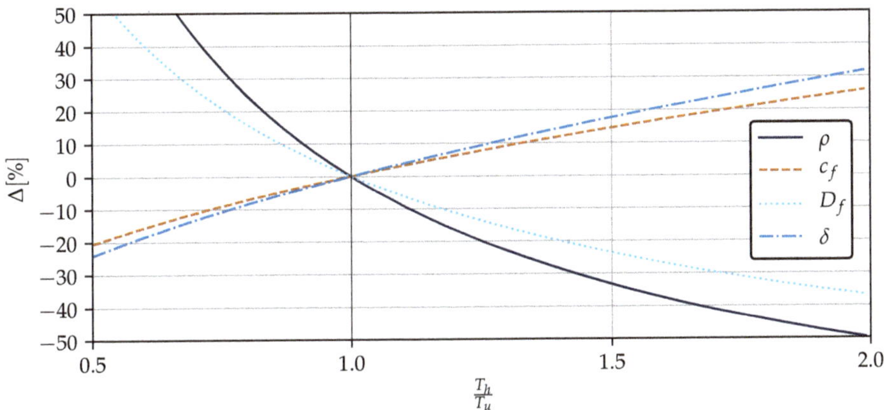

Figure 6. Theoretical impact of wall heating/cooling on a smooth flat plate turbulent boundary layer density, skin friction coefficient, skin friction drag force and boundary layer 99% thickness compared to an unheated wall. Valid for $Re_x = 10^6 - 10^8$.

Wall heating not only has an impact on skin friction drag but also effects the pressure drag. The turbulent boundary layer velocity profile thickens, because [34]:

$$\delta = \frac{0.37x}{Re_x^{\frac{1}{5}}}. \tag{14}$$

For a flat plate, the pressure gradient is zero at all locations. For a slender body (fuselage) or lifting surface (wing, tail planes), however, the pressure gradient varies in stream wise direction. Therefore, for a three-dimensional curved body, the heating of the wall has an effect on the (not-separated) pressure drag as shown by Lin and Ash. The heating of the wall increases the turbulent displacement thickness (δ^*), which in turn leads to a slight increase in pressure drag [32]. In addition, the boundary layer shape factor is increased. Thus, the adverse pressure gradient is increased, causing an earlier flow separation [32,35]. The effect of wall heating on pressure drag is small compared to the effect on skin friction drag [32].

In summary, in regions in which the boundary layer is laminar, an increased temperature leads to an earlier transition and, thus, to an increase in total D_f. To make use of the beneficial effect of wall heating on the turbulent drag force, the surface of aircraft components should preferably be heated only in regions in which a fully turbulent boundary layer is present. This means that for example, the fuselage nose (cockpit area) or wing leading edge (slats etc.) should not be used for heat disposal. For aircraft concepts that unite different technologies, which emit excessive heat and aim at an increased laminar flow control, detailed studies have to be conducted, compromising excessive heat disposal and drag reduction approaches.

4. Surface Cooling Potential

For the following studies the simple correlations derived in Sections 2 and 3 are combined to estimate Q_{av} depending on *MTOW*. The calculated A_{wet} is reduced for each component to account for more realistic cases with unusable surface area in each component. These reductions are based on observations and estimations from drawings in manufacturer's documents such as in References [18,19].

4.1. Area Reduction Assumptions

In Section 3.3 it was shown that heating surfaces underneath laminar flow has a negative effect on aircraft drag. Therefore, areas at the front of each component are avoided as locations for surface heat exchangers. Independent of the size of the aircraft, the first 4 m of the fuselage are not used because cockpit, sensors and nose landing gear bay are located here. In addition, the contraction of this part is responsible for the overestimation of A_{wet} of the fuselage in Section 2. The rear 15% of the fuselage length are also not used because of the tail plane attachments, the auxiliary power unit (APU) and again the contraction that lead to an overestimation of A_{wet}. For the remaining fuselage middle section, a stripe of 0.5 m width is spared on both sides to account for the windows. Another 10% is subtracted from the total middle section area to account for passenger doors, cargo doors and landing gear doors as well as sensors and air openings. The wing leading edge and trailing edge (20% chord length each) cannot be used as a heat sink due to slats, flaps and other control surfaces. Only the forward 50% of the nacelle length is used to account for possible thrust reversers installed in the back. The rear 33% of the horizontal and vertical tail plane's chord length are not used because of the installed control surfaces. In the following, all remaining surfaces are employed for heat rejection.

4.2. Cooling Potential for Typical Operating Points

Q_{av} is investigated in multiple typical operating points: TO, HTO, CL and Cruise CR. The atmospheric conditions (*Ma*, *alt* and dT_{ISA}) of each operating point are listed in Table 5. The design space includes *MTOW* over the entire range of the database used in Section 2.1 as well as T_{surf} ranging from 320 K to 400 K. T_{surf} has to be lower than the maximum allowed operating temperature of electric components, which for motors can be up to 180 °C [36] but are significantly lower for batteries. The actual T_{surf} depends on the installed drive train and the hot side of the cooling system. This study shows Q_{av} for a wide range of T_{surf} in Figure 7.

Table 5. Investigated operating points.

Opearting Point	Ma (-)	alt (m)	dT_{ISA} (K)
TO	0.2	0	0
HTO	0.2	0	+20
CL	0.5	5000	0
CR	0.8	10,000	0

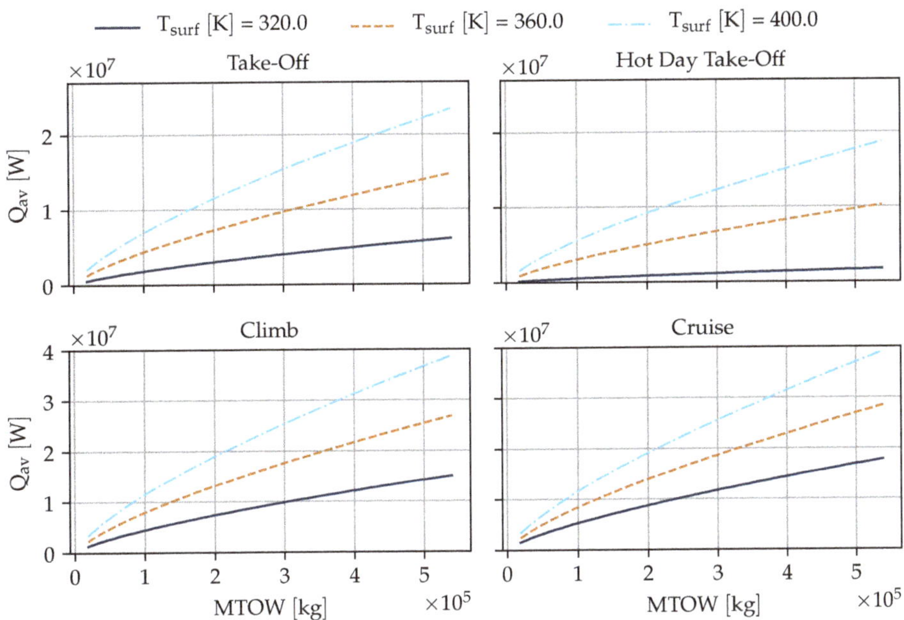

Figure 7. Q_{av} in multiple operating points for aircraft equipped with surface heat exchangers.

Figure 7 can be used to estimate Q_{av} of any tube and wing aircraft with known *MTOW*. For example, the A320 sized aircraft from Section 2.2 has an *MTOW* of 71,000 kg. Assuming T_{surf} of 360 K an estimated Q_{av} of approximately 250 kW in HTO—the most critical condition—results. In contrast, the same aircraft with the same T_{surf} would be able to reject about 7 MW of heat in CR. In all operating points, Q_{av} increases with *MTOW*, because A_{wet} increases. The slope of Q_{av} decreases with *MTOW*, because the *MTOW*–A_{wet} correlation is weakly logarithmic and because aircraft with higher *MTOW* have increased flow lengths on all surfaces, which results in lower α_x towards their rear ends (cf. Section 3.2). Q_{av} also increases with increasing T_{surf} due to the higher temperature difference to the ambient. In the HTO case, Q_{av} is about five times as large for $T_{surf} = 360$ K than the value corresponding with $T_{surf} = 320$ K over the entire *MTOW* range. The high sensitivity is due to a relatively high ambient temperature (T_{amb}), resulting in an increase of heat transfer driving temperature difference (ΔT) from $\Delta 12$ K to $\Delta 52$ K (roughly factor five). For CL and CR the relative T_{surf} sensitivity is not as strong because T_{amb} is lower. Over the entire *MTOW* range, Q_{av} is about twice as large for CR and CL compared to TO. The ratio even increases when comparing CR and CL to HTO. The reasons for this large difference are the lower T_{amb} in CL and CR compared to TO and HTO as well as the higher *Ma* that increases convection. The difference in Q_{av} between CL and CR is approximately 10% over the entire *MTOW* range for $T_{surf} = 320$ K. The difference is less for higher T_{surf} and hardly noticeable for the largest T_{surf} of 400 K. CR has a lower T_{amb} than CL which results in a larger ΔT. The relative difference between ΔT_{CL} and ΔT_{CR} decreases with increasing T_{surf}. Also, for heat transfer, the total T_{amb} is relevant and due to the increased *Ma* in CR it is not smaller by the same ratio compared to CL as the static T_{amb}. The increased flight speed should additionally result in higher *Nu* in CR but the effect is reduced by the lower ρ_{amb}. More elaborate studies on the dependence of forced convection on flight conditions can be found in [16].

4.3. Hot Day Take-Off Performance

Results from the previous section indicate that HTO is the condition with minimum Q_{av}. Additionally, the propulsive power is usually at its maximum during TO, which means Q_{rq} is at its maximum as well.

For an aircraft application, the most relevant metric is the ratio of Q_{av} to Q_{rq} (C_Q). The following study is conducted for HTO conditions. Q_{rq} differs depending on the electric architecture and the mission profile amongst others. For the first part of this study a fully power hybridized aircraft ($H_P = 1$) is assumed with estimated values for the efficiencies required by the methods shown in Section 2.3 listed in Table 6. The effect of varying $MTOW$ over the entire range of the database (cf. Section 2.1) as well as T_{surf} ranging from 320K to 400K is investigated. The results are shown in Figure 8a. For the second part of the study different H_P values are assumed and the required T_{surf} to achieve $C_Q = 1$ that is, a Q_{av} that matches Q_{rq} is investigated. Figure 8b shows the results.

Table 6. Values for the estimation of Q_{rq}.

η_{trans}	0.5
η_{ec}	0.9
H_P	1.0
Ma_{TO}	0.2

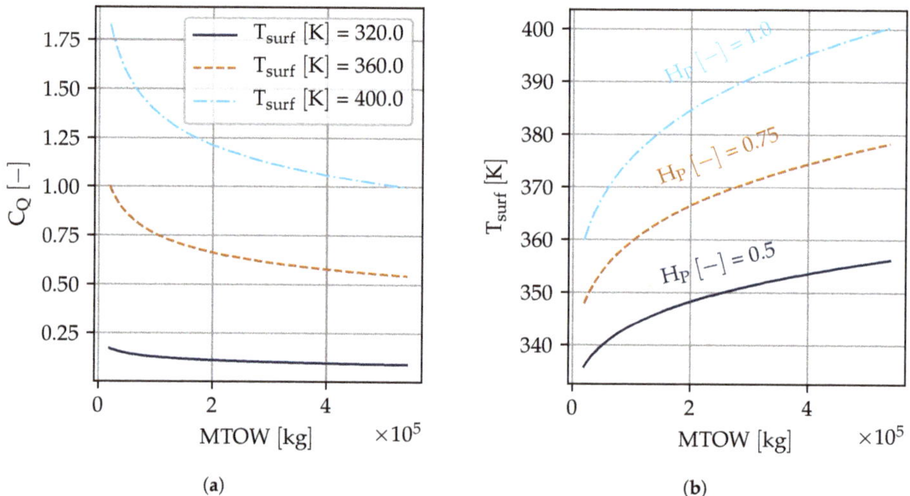

Figure 8. Comparison of Q_{av} and Q_{rq} for hybrid electric aircraft in hot day take-off conditions. (**a**) Ratio of Q_{av} to Q_{rq} for different T_{surf}. (**b**) Required T_{surf} to achieve $C_Q = 1$.

A first observation is that Q_{av} and Q_{rq} are within the same order of magnitude during HTO. Within the used parameter ranges values above and below unity exist for C_Q. C_Q is decreasing linearly with T_{surf} and hyperbolically with $MTOW$. Smaller aircraft have a favourable C_Q. This is mainly due to the increased flow length on all surfaces of larger aircraft but also due to the weakly logarithmic behaviour of the $MTOW$–A_{wet} correlation. For the smallest considered aircraft, C_Q ranges between 0.2 and 1.8 depending on T_{surf}. In contrast, the C_Q range for the largest considered aircraft is between 0.1 and 1.0. In Figure 8b the required T_{surf} during HTO to achieve $C_Q = 1$ is depicted for different H_P. T_{surf} grows linearly with H_P because Q_{rq} increases proportionally to H_P. T_{surf} grows logarithmically with $MTOW$, which is expected from Figure 8a: Smaller aircraft have an advantage over large aircraft with regards to potential cooling via existing aircraft surfaces. Taking the A320 sized aircraft from Section 2.1 ($MTOW = 71,000$ kg) as an example again with $H_P = 1$, Figure 8b shows that T_{surf} of about 370 K would be required during HTO to provide enough cooling power for the waste heat load of the drive train. Heating up the surface to an average T_{surf} of 370 K is going to be challenging in an application with low grade waste heat potentially involving artificial measures such as vapour compression cycles to increase the temperature at which heat is rejected. However, such systems add weight and need power, which might diminish the benefits from a surface cooling system on aircraft level. A comparison of a conventional cooling system with a surface cooling system on aircraft level will be performed in future studies. The results shown in Figure 8 are for steady state cooling in the most adverse conditions. A dynamic model might reveal that requiring steady state cooling during TO is unnecessary because thermal inertia of components and fluids can cope with temporarily high heat loads that is, $Q_{rq} > Q_{av}$. The dynamic behaviour of surface cooling systems will also be part of future work. The feasibility of using surface heat exchangers for cooling highly depends on Q_{rq} and the requirements for T_{surf} are more relaxed for $H_P < 1$. For the aforementioned example reducing H_P to 0.5 results in a 30 K decrease in required T_{surf} to about 340 K. Thus, surface cooling might be a viable option for aircraft with lower H_P or can be used in combination with a conventional cooling system for aircraft with large electrification to reduce heat exchanger size and drag.

5. Conclusions and Outlook

The potential of using the existing aircraft surfaces as heat sink for the waste heat of a (hybrid-) electric drive train was investigated. First, empirical correlations were derived to predict an aircraft's wetted area (A_{wet}) from its maximum take-off weight ($MTOW$). The database included aircraft ranging from small regional aircraft to large twin aisle aircraft. The chosen correlation was a fit of the log-log scaled data that had a coefficient of determination (r^2) of 0.986. To assess the ratio of available cooling power to required cooling power (C_Q), a simple estimation of the waste heat based on take-off thrust was used. Heat transfer from wetted surfaces was modelled via flat plate correlations. To apply them, the total A_{wet} was divided into five component groups: fuselage, wing, nacelles, horizontal tail and vertical tail. The mean of the relative area share was calculated for each component.

Sensitivities of the heat transfer model were studied. The flow transition had a considerable impact on the predicted heat flow. The applied methods in this work did not include accurate transition prediction. The detailed analysis of the heat transfer potential of surfaces with large laminar shares are part of future work while the results of this study may be used for concepts where turbulent flow dominates. Other sensitivities investigated were wing taper ratio and aspect ratio as well as fuselage slenderness ratio. Their impact was too small to be further considered because it was below the expected uncertainty level from the modelling simplifications. A qualitative assessment of the impact of surface heating on the aircraft's drag was performed. When heat is added to a laminar flow region an increase in skin friction drag is expected. The opposite is true for fully turbulent flow regions where heat addition reduces skin friction drag. A quantification of the expected effects is part of future work. Combining the findings for the drag with the flow transition sensitivity of the heat transfer leads to the conclusion that surface heat exchangers should only be installed in fully turbulent flow regimes to avoid a negative impact of surface heating on the aircraft aerodynamics.

Additional area reductions to account for unusable surface area for example, windows, landing gear doors and cockpit were applied and available cooling power (Q_{av}) were calculated for a range of $MTOW$ over the entire database and average surface temperatures (T_{surf}) between 320 K and 400 K. Q_{av} was evaluated in four operating points: Take-off (TO), Hot Day Take-off (HTO), Climb (CL) and Cruise (CR). Q_{av} was largest in CR with about 7 MW for an A320 size aircraft and a medium T_{surf} of 360 K. The most critical operating point was HTO with Q_{av} of only 0.25 MW for the aforementioned aircraft and T_{surf}.

C_Q was calculated in HTO. The smallest aircraft showed an advantage over larger aircraft with C_Q values ranging from 0.2 to 1.8 depending on T_{surf} compared to 0.1 to 1.0 for the largest aircraft.

The results of this study may be used to quickly assess the feasibility of a surface cooling concept for a (hybrid-) electric aircraft. Future work will include more detailed models for surface heat transfer. Instead of assuming an average T_{surf}, surface heat exchangers with a hot side flow will be modelled. These models that can also be used in a dynamic simulation will allow a more detailed sizing of the thermal management system. To assess heat transfer more precisely in a 3D flow field, numerical methods will be developed Those methods may go beyond the scope of a conceptual aircraft analysis and are part of more in-depth studies later in the design process.

The impact of adding heat to the boundary layer has only been qualitatively assessed in this work. Numerical methods will help to quantify the effect. Together with improved drag predictions, mass and power estimations the concept will be compared to a similar aircraft with a conventional cooling system to quantify its benefits. In addition, structural integration of surface heat exchangers may be a challenge. The concept will be evaluated with regard to maintainability.

Author Contributions: Conceptualization, methodology, simulation, analysis and writing of all aspects of the research except for Section 3.3, H.K.; Conceptualization, methodology, simulation, analysis and writing of Section 3.3, A.L.H.; Supervision, M.H. All authors have read and agreed to the published version of the manuscript.

Funding: This research received funding as part of SynergIE, a research project supported by the Federal Ministry for Economic Affairs and Energy in the national LuFo V program. Any opinions, findings and conclusions expressed in this document are those of the authors and do not necessarily reflect the views of the other project partners.

Acknowledgments: The authors would like to thank Arne Seitz for his continued support of the research, his critical review and many fruitful discussions.

Conflicts of Interest: The authors declare no conflict of interest. The funders had no role in the design of the study; in the collection, analyses, or interpretation of data; in the writing of the manuscript, or in the decision to publish the results.

Abbreviations

SRIA	Strategic Research and Innovation Agenda
TMS	Thermal Management System
ISA	International Standard Atmosphere
TO	Take-off
HTO	Hot Day Take-off
CL	Climb
CR	Cruise
ACOC	Air Cooled Oil Cooler
SACOC	Surface Air Cooled Oil Cooler
APU	Auxilliary Power Unit
MTOW	Maximum Take-off Weight
MPL	Maximum Payload

Roman Symbols

A	Area
dT	Temperature deviation
T	Temperature
alt	Altitude
Ma	Mach number
Q	Heat rate
n	Number
R	Range
r^2	Coefficient of determination
H	Degree of Hybrdization
P	Power
F	Thrust
v	Velocity
Nu	Nusselt Number
Re	Reynolds Number
AR	Aspect Ratio
l	Length
d	Diameter
x	Coordinate in flow direction
c	Coefficient
D	Drag Force
TR	Temperature Ratio
C_Q	Ratio of Heat Rates

Greek Symbols

σ	Standard deviation
δ	Boundary layer thickness
Δ	Difference
η	Efficiency
α	Heat transfer coefficient
λ	Wing taper ratio
Λ	Slenderness ratio
ρ	Density
μ	Dynamic Viscosity

Subscripts

wet	wetted
rq	required
exp	exposed
sim	simplified
act	actual
max	maximum
des	design
tot	total
av	available
trans	transmission
ec	electric
rad	radiation
surf	surface
c	critical
min	minimum
f	friction
h	heated
u	unheated
amb	ambient

References

1. Advisory Council for Aviation Research and Innovation in Europe. *Strategic Research and Innovation Agenda: Volume 1: 2017 Update*; Advisory Council for Aviation Research and Innovation in Europe: Brussels, Belgium, 2017.
2. Yakinthos, K.; Donnerhack, S.; Misirlis, D.; Flouros, M.; Vlahostergios, Z.; Goulas, A. Intercooled Recuperated Aero Engine: Early Development Stages and Optimization of Recuperation Based on Conventional Heat Exchangers. In Proceedings of the 2nd ECATS Conference, Athens, Greece, 7–9 November 2016.
3. Kaiser, S.; Kellermann, H.; Nickl, M.; Seitz, A. A Composite Cycle Engine Concept for Year 2050. In Proceedings of the 31st Congress of the International Council of the Aeronautical Sciences, Belo Horizonte, Brazil, 9–14 September 2018.
4. Jacob, F.; Rolt, A.; Sebastiampillai, J.; Sethi, V.; Belmonte, M.; Cobas, P. Performance of a Supercritical CO_2 Bottoming Cycle for Aero Applications. *Appl. Sci.* **2017**, *7*, 255. [CrossRef]
5. Rosero, J.A.; Ortega, J.A.; Aldabas, E.; Romeral, L. Moving towards a more electric aircraft. *IEEE Aerosp. Electron. Syst. Mag.* **2007**, *22*, 3–9. [CrossRef]
6. Pornet, C.; Isikveren, A.T. Conceptual design of hybrid-electric transport aircraft. *Prog. Aerosp. Sci.* **2015**, *79*, 114–135. [CrossRef]
7. Schlabe, D.; Lienig, J. *Model-Based Thermal Management Functions for Aircraft Systems*; SAE Technical Paper Series; SAE International: Warrendale, PA, USA, 2014. [CrossRef]
8. Wang, T.; Britcher, C.; Martin, P. Surface heat exchangers for aircraft applications—A technical review and historical survey. In Proceedings of the 37th Aerospace Sciences Meeting and Exhibit, Reno, NV, USA, 11–14 January 1999; American Institute of Aeronautics and Astronautics: Reston, VA, USA, 1999; p. 245. [CrossRef]

9. Wilkinson, S.P. Interactive wall turbulence control. In *Viscous Drag Reduction in Boundary Layers*; Bushnell, D.M., Hefner, J.N., Eds.; Progress in Astronautics and Aeronautics; American Institute of Aeronautics and Astronautics: Reston, VA, USA, 1990.

10. Sousa, J.; Villafañe, L.; Paniagua, G. Thermal analysis and modeling of surface heat exchangers operating in the transonic regime. *Energy* **2014**, *64*, 961–969. [CrossRef]

11. European Commission. *Surface Heat Exchangers for Aero-Engines*; European Commission: Brussels, Belgium, 2016.

12. European Commission. *Surface Heat Exchangers For Aero Engines 2*; European Commission: Brussels, Belgium, 2019.

13. European Commission. *Aerodynamic upgrade of Surface Air Cooled Oil Cooler*; European Commission: Brussels, Belgium, 2019.

14. Sakuma, Y.; Saito, H.; Watanabe, T.; Himeno, T.; Inoue, C.; Tomida, S.; Takahashi, N. Conjugate Heat Tansfer Analysis on Plate Fin Surface Air Cooled Oil Cooler. In Proceedings of the Shanghai 2017 Global Power and Propulsion Forum, Shanghai, China, 30 October–1 November 2017.

15. Liu, J.; Peck, J.; Yazawa, K.; Fisher, T.S.; Shih, T.I.P. Bypass, Loss, and Heat Transfer in Aircraft Surface Coolers. *Front. Mech. Eng.* **2019**, *5*, 219. [CrossRef]

16. Kellermann, H.; Habermann, A.L.; Vratny, P.C.; Hornung, M. Assessment of Fuel as Alternative Heat Sink for Future Aircraft. *Appl. Therm. Eng.* **2019**, under review.

17. Jenkinson, L.; Simpkin, P.; Rhodes, D. *Civil Jet Aircraft Design*; American Institute of Aeronautics and Astronautics, Inc.: Washington, DC, USA, 1999. [CrossRef]

18. Boeing Commercial Airplanes. *747-8 Airplane Characteristics for Airport Planning*; Boeing Commercial Airplanes: Seattle, WA, USA, 2012.

19. Airbus S.A.S. *A350 Aircraft Characteristics Airport and Maintenance Planning*; Airbus S.A.S.: Blagnac, France, 2016.

20. Torenbeek, E. *Synthesis of Subsonic Airplane Design: An introduction to the Preliminary Design of Subsonic General Aviation and Transport Aircraft, with Emphasis on Layout, Aerodynamic Design, Propulsion and Performance*; Springer: Dordrecht, The Netherlands, 2010.

21. Pornet, C.; Gologan, C.; Vratny, P.C.; Seitz, A.; Schmitz, O.; Isikveren, A.T.; Hornung, M. Methodology for Sizing and Performance Assessment of Hybrid Energy Aircraft. In Proceedings of the 2013 Aviation Technology, Integration, and Operations Conference, Los Angeles, CA, USA, 12–14 August 2013; American Institute of Aeronautics and Astronautics: Reston, VA, USA, 2013; p. 35. [CrossRef]

22. Lorenz, L.; Seitz, A.; Kuhn, H.; Sizmann, A. Hybrid Power Trains for Future Mobility. In Proceedings of the 62. Deutscher Luft- und Raumfahrtkongress, Stuttgart, Germany, 10–12 September 2013; Deutsche Gesellschaft für Luft- und Raumfahrt: Bonn, Germany, 2013.

23. Haaland, S.E. Simple and Explicit Formulas for the Friction Factor in Turbulent Pipe Flow. *J. Fluids Eng.* **1983**, *105*, 89. [CrossRef]

24. Incropera, F.P.; DeWitt, D.P.; Bergman, T.L.; Lavine, A.S. *Principles of Heat and Mass Transfer*, 7th ed.; International Student Version; Wiley: Singapore, 2013.

25. Schultz-Grunow, F. New frictional resistance law for smooth plates. *Luftfahrtforschung* **1940**, *1940*, 239–246.

26. Henninger, J.H. *Solar Absorptance and Thermal Emittance of Some Common Spacecraft Thermal-Control Coatings*; NASA Goddard Space Flight Center: Greenbelt, MD, USA, 1984.

27. Arnal, D.; Reneaux, J.; Casalis, G. Numerical and Experimental Studies Related to Skin Friction Drag Reduction Problems. Proceedings of the Colloquium Transitional Boundary Layers in Aeronautics. 1996. Available online: https://www.dwc.knaw.nl/DL/publications/PU00011219.pdf (accessed on 19 November 2019).

28. Reshotko, E. Drag Reduction by Cooling in Hydrogen-Fueled Aircraft. *J. Aircr.* **1979**, *16*, 584–590. [CrossRef]

29. Masad, J.A.; Nayfeh, A.H. Laminar flow control of subsonic boundary layers by suction and heat–transfer strips. *Phys. Fluids A Fluid Dyn.* **1992**, *4*, 1259–1272. [CrossRef]

30. Dovgal, A.V.; Levchenko, V.; Timopeev, V.A. Boundary Layer Control by a Local Heating of the Wall. In *Laminar-Turbulent Transition*; Springer: Berlin/Heidelberg, Germany, 1990; pp. 113–121.

31. Kramer, B.; Smith, B.; Heid, J.; Noffz, G.; Richwine, D.; Ng, T. Drag reduction experiments using boundary layer heating. In Proceedings of the 37th Aerospace Sciences Meeting and Exhibit, Reno, NV, USA, 11–14 January 1999. [CrossRef]

32. Lin, J.C.; Ash, R. Wall temperature control of low-speed body drag. *J. Aircr.* **1986**, *23*, 93–94. [CrossRef]

33. Hoerner, S.F. *Fluid-Dynamic Drag: Theoretical, Experimental and Statistical Information*; Hoerner Fluid Dynamics: Bakersfield, CA, USA, 1965.
34. Schlichting, H.; Gersten, K. *Boundary-Layer Theory*; Springer: Berlin/Heidelberg, Germany, 1979. [CrossRef]
35. Spalart, P.R.; Watmuff, J.H. Experimental and numerical study of a turbulent boundary layer with pressure gradients. *J. Fluid Mech.* **1993**, *249*, 337–371. [CrossRef]
36. Ohta, S. Temperature Classes of Electrical Insulators. *Three Bond Technical News*, 1 December 1985.

Article

Determination of Serviceability Limits of a Turboshaft Engine by the Criterion of Blade Natural Frequency and Stall Margin

Yaroslav Dvirnyk [1,2]**, Dmytro Pavlenko** [2,3] **and Radoslaw Przysowa** [4,5,*]

1   Motor Sich JSC, 15 Motorostroiteley Ave.,69068 Zaporizhzhia, Ukraine; dvirnyk@gmail.com
2   Mechanical Engineering Department, National University "Zaporizhzhia Polytechnic", 64 Zhukovskogo st, 69061 Zaporizhzhia, Ukraine; dvp1977dvp@gmail.com
3   Zaporizhzhia Machine-Building Design Bureau Progress State Enterprise Named After Academician A.G. Ivchenko, 2 Ivanova st, 69068 Zaporizhzhia, Ukraine
4   Instytut Techniczny Wojsk Lotniczych, ul. Księcia Bolesława 6, 01-494 Warsaw, Poland
5   Technology Partners Foundation, ul. Pawińskiego 5A, 02-106 Warsaw, Poland
*   Correspondence: radoslaw.przysowa@itwl.pl

Received: 2 November 2019; Accepted: 4 December 2019; Published: 9 December 2019

Abstract: This paper analyzes the health and performance of the 12-stage axial compressor of the TV3-117VM/VMA turboshaft operated in a desert environment. The results of the dimensional control of 4800 worn blades are analyzed to model the wear process. Operational experience and two-phase flow simulations are used to assess the effectiveness of an inlet particle separator. Numerical modal analysis is performed to generate the Campbell diagram of the worn blades and identify resonant blade vibrations which can lead to high cycle fatigue (HCF): mode 7 engine order 30 in the first stage and mode 8 engine order 60 in the fourth. It is also shown that the gradual loss of the stall margin over time determines the serviceability limits of compressor blades. In particular, the chord wear of sixth-stage blades as high as 6.19 mm results in a reduction of the stall margin by 15–17% and a permanent stall at 770–790 flight hours. In addition, recommendations setting out go/no-go criteria are made to maintenance and repair organizations.

Keywords: gas-turbine performance; turboshaft; axial compressor; blade; FEM; CFD; erosion; wear; stall margin; compressor surge; brownout

1. Introduction

Long-term engine performance in adverse operating conditions is one of the main prerequisites for successful helicopter missions. Unlike airplanes, helicopters have to hover for a long time above the ground, often raising a cloud of dust and causing a brownout. Therefore, the ability to ensure long-term operations at elevated dust concentrations is one of their most important features. Studies on the impact of environmental particles on the efficiency of helicopter engines were already being carried out during the development of the first helicopter types [1].

The gas path of the helicopter engine operated in a dusty environment is exposed to contamination [2,3]. Deposition of particles on aerofoils [4,5] and the heat transfer surfaces of the air cooling system [6] can lead to the deterioration of the aerodynamic and thermodynamic properties of these components. Moreover, given that dust particles (no matter how small they are) have abrasive properties, they also cause erosive wear of the engine components and contribute to various types of structural damage [7–9]. Erosive wear of compressor blades leads to reduction in the stall margin of the compressor, an increase in the likelihood of fatigue damage of compressor blades due to changes in their natural frequencies of vibration, and a decrease in the efficiency of the engine due to the wear of the gas path.

Rotating components are sensitive to mechanical damage caused by solid particles [10]. Models describing their impact and related erosive damage have been proposed by many researchers, such as Finney [11], Bitter [12], and Sheldon [13], but predicting the actual aerofoil degradation is still difficult. The main factors affecting the magnitude of erosion consist of the angle of collision, velocity and particle size, blade surface properties, and particle concentration. Van der Walt [14] showed that the wear rate of the aerofoils is directly proportional to dust concentration. The mechanical properties of the material are also an important parameter influencing the mechanisms of erosion [15,16].

Eustyfeev [17] performed experimental and theoretical studies of compressor blades eroded by solids to determine the erosion resistance of the blade material EI-961. The limiting ratio of the particle size to their velocity of contact with the material was determined. However, the experiment was performed on a cut-out sample, not on a real blade, as well as on one type of material, that is EI-961 steel, while most blades of modern compressors are made of titanium alloy.

Batcho [18] and Singh [19] analyzed loss of performance caused both by erosion and deposition. The impact of ingested particles on engine operation depends on the physical and chemical properties of the dust, its composition and concentration. Particles deposited on compressor aerofoils change their geometry and roughness, which leads to a decrease in the efficiency, and consequently, reduces the pressure ratio and performance of the compressor [20].

The time between overhauls (TBO) of an engine operated in a highly dusty environment is much less than that set out by the manufacturer and is reduced by the erosion of the compressor blades. The statistical analysis [21] showed that engines with eroded compressors represent the largest group of engines grounded due to compressor blade damage (30–35%). This value is comparable to the ratio of engine removals due to foreign objects ingested from the runway during take-off (25–30%). The share of aviation incidents, bird ingestion, or human errors during maintenance accounts for 15–20%.

The aim of this study was to assess the serviceability limits of the compressor blades of a helicopter engine operating in a dusty environment. To achieve this goal, the following tasks based on the results of the dimensional control of worn blades were performed:

- Establishing patterns of blade wear as a function of flight hours (FH) and dust concentration;
- Evaluating the increase in the natural frequency of blades by modeling the geometry of worn aerofoils over the engine operating time;
- Development of a methodology for modeling the flow through the axial compressor;
- Calculating the compressor maps describing the blades with different degrees of wear.

2. Methods

The TV3-117 turboshaft (Figure 1) with a maximum power of 1640 kW and overall pressure ratio of 9.4 belongs to one of the world's most popular helicopter engines. It rotates anticlockwise (looking from aft to fore) and has a 12-stage axial compressor with a subsonic inlet and variable inlet guide vanes. Close to the blade tip, the Mach number is M = 0.8–0.92. The tip radius is 163 mm. The number of blades in subsequent stages are as follows: 37, 43, 59, 67, 73, 81, and 89 in further stages.

This work aims to analyze and predict the erosive wear of the TV3-117 engines powering the Mi-8MTV and Mi-24 helicopters in the desert environment of the Republic of Algeria. The region of North Africa and the Middle East is known for extreme ambient temperatures, a high concentration of dust, and sand dispersing up to a height of 6000 m, which is the helicopter operation zone. For example, in North Sudan, dust concentration is $1.3\,\mathrm{g/m^3}$, and in Algeria $1.3–1.6\,\mathrm{g/m^3}$. The size of particles ingested into the gas path ranges from 0.01 to 2 mm.

The operation of gas-turbine engines in the regions described above, with a high content of dust and sand, inevitably leads to deterioration of the engine [22,23]. Airborne particles may cause substantial erosive wear of compressor blades. As a consequence, tip clearance is widened and the aerofoil profile is modified—in particular, the chord length and the blade thickness appear to decrease, especially above 66% of the span (Figure 2). This leads to a reduction in the compressor pressure ratio and ultimately a reduction in the efficiency of the entire engine.

Firstly, a statistical analysis was carried out to study the nature of compressor wear and to determine its critical components. On the basis of the subsequent regression analysis, patterns of wear of the blades of all compressor stages as a function of engine operation time and dust concentration were established.

The obtained patterns of the chord wear of blades of all compressor stages were used to predict their geometry, depending on the engine operating time. This allowed for the simulation of the flow through the worn compressor and modeling blade vibration, and a determination of the serviceability limit of the blades in terms of structural integrity and stall margin.

For the analyzed engines, the observed impact of erosion on compressor performance was several times greater than deposition due to high dust concentration and fair inlet protection. Moreover, compressor fouling can be reversed by washing. Therefore, particle deposition is not taken into account in this work.

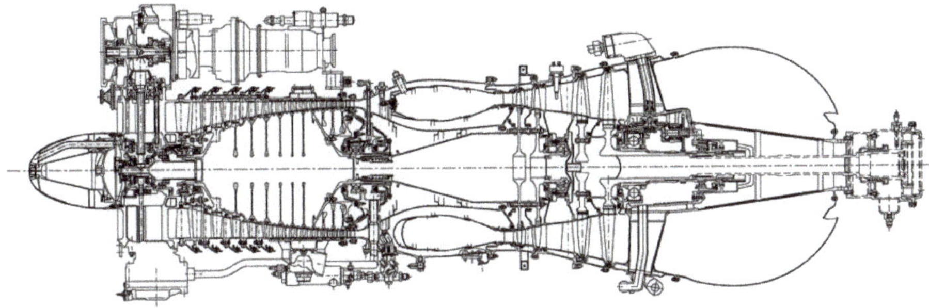

Figure 1. TV3-117 turboshaft [24].

Figure 2. Erosive wear of compressor blades.

2.1. Blade Inspection

The inspected engines were grounded and torn down for one of two reasons: Either due to the chord wear of first-stage blades higher than 2 mm at the tip; or due to exceeding the threshold levels of performance parameters such as the gas generator speed n_1 or the turbine inlet temperature (TIT).

The dimensional control of blade geometry was performed by measuring the chord and the thickness of the profile at various sections as well as blade height. Ten blades were selected for the evaluation from each stage of 40 engines [25]. In total, 4800 blades were inspected. Statistica 12 was used to analyze the results.

The inspected engines were initially divided into two groups. The first one included those that were operated without a particle separator (IPS), and the second with an IPS.

2.2. Structural Analysis

Compressor blades of modern turboshafts are characterized by thin profiles and relatively low stiffness, hence forced vibration presents a notable threat to them [26,27]. Compressor blades absorb intensive static and dynamic loads. When rotating under large centrifugal forces, the blades are deformed, which leads among others to a decrease in their twist. As a result of unsteady flow forces, compressor blades vibrate relative to their static deformations. The distribution of dynamic stresses in blades has to be determined to ensure that they remain below the fatigue limit of the material [28,29].

The erosive wear of blades has a significant impact on their strength, which was confirmed by Hamed [1,30]. The properties of the blade surface and its ability to withstand erosion are one of the key factors that determine the reliability of the system, since a variety of surface irregularities can lead to stress concentration that increases the risk of high cycle fatigue (HCF) [31–34].

The main reason for forcing blade vibration is an irregularity of flow in the circumferential and radial direction. The frequencies of the driving forces are multiples of the rotor speed (engine orders) that are equal to the number of obstacles around the circumference, e.g., the number of guide vanes and struts in the gas path. When the speed changes, a number of resonant frequencies can be excited. To predict the frequencies of resonant vibration, a modal analysis of blades and the Campbell diagram are necessary [35,36].

Vibration frequencies and operational deflection shapes can be determined with a certain degree of accuracy by numerical methods, in particular using volumetric finite element models (FEM). At present, this research method is preferred, since the analytical method of calculating frequencies and vibration modes is not suitable for the complex geometry of aerofoils.

Blade geometry was measured using 3D scanning, followed by postprocessing in the CAD system ASCON KOMPAS 16.0. Geometry models of compressor blades were developed in Unigraphics NX8. Compressor blades operated in a dusty environment were described using parameterized solid-state models. By varying the blades height, chord length, and thickness in the corresponding sections of the aerofoil, several variants of blade geometry were generated.

The grid models of blades, created with the ANSYS ICEM CFD grid generator, consisted of 15–18 thousand hexagonal SOLID 185 elements. The natural frequencies of blades were calculated by the ANSYS 14.5 solver.

2.3. CFD Model

The 3D flow calculation, with an ideal gas as the working fluid, was based on the Navier–Stokes equations and the finite element method (FEM), implemented in the ANSYS CFX solver. The mesh was created in ANSYS Turbo Grid to model the operation of the compressor (Figure 3). A separate mesh flow model (domain) was designed for each stage (Figures 4 and 5). The domains were designed taking into account the possibility of air flow leaking in the radial gap (Figure 6) by adding the interface in the blade tip. The compressor model consisted of 26 domains (Table 1).

The following grid parameters were considered when building the grid:

- ATM optimized topology ensures a high-quality mesh with hexahedral elements for twisted aerofoils;
- Parameter $y\pm$ size of the first wall element has a value within (80–160) units;
- The ratio of the dimensions of the elements does not exceed 6.

First, the flow calculation was performed for two rotational speeds: 95% and 98%, where 1% corresponds to 195.37 rpm. An ambient air temperature of 288 K and pressure of 101.325 kPa were assumed. Each rotational speed of the compressor rotor corresponds to certain angles of the variable inlet guide vanes (VIGV) and guided vanes of the further four stages.

Table 1. Computational fluid dynamics (CFD) domain and mesh parameters.

Domain	Number of Nodes	Number of Elements
VIGV	721,686	687,199
Rotor 1	809,883	768,068
Stator 1	708,708	664,875
Rotor 2	807,087	768,737
Stator 2	695,730	646,335
Rotor 3	801,327	748,912
Stator 3	682,752	643,632
Rotor 4	796,419	757,004
Stator 4	669,777	630,693
Rotor 5	791,046	739,303
Stator 5	656,799	612,532
Rotor 6	783,924	744,989
Stator 6	643,821	608,066
Rotor 7	768,588	727,334
Stator 7	630,843	599,284
Rotor 8	765,987	724,804
Stator 8	617,865	573,627
Rotor 9	762,795	725,798
Stator 9	604,887	574,900
Rotor 10	761,880	715,822
Stator 10	591,909	549,479
Rotor 11	754,461	703,549
Stator 11	578,934	543,629
Rotor 12	753,300	710,338
Deswirl 1	565,956	536,180
Deswirl 2	559,467	522,532
Total	18,285,831	17,227,621

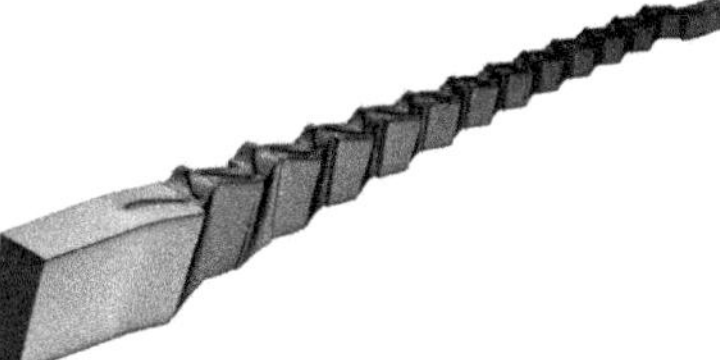

Figure 3. CFD model of the compressor.

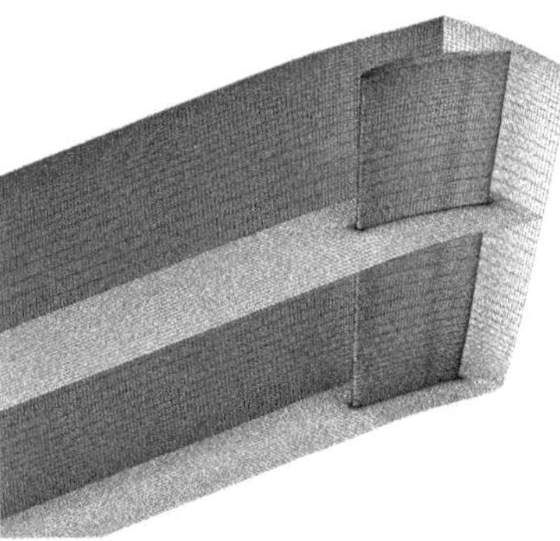

Figure 4. Mesh of the flow through the inlet and variable inlet guide vanes.

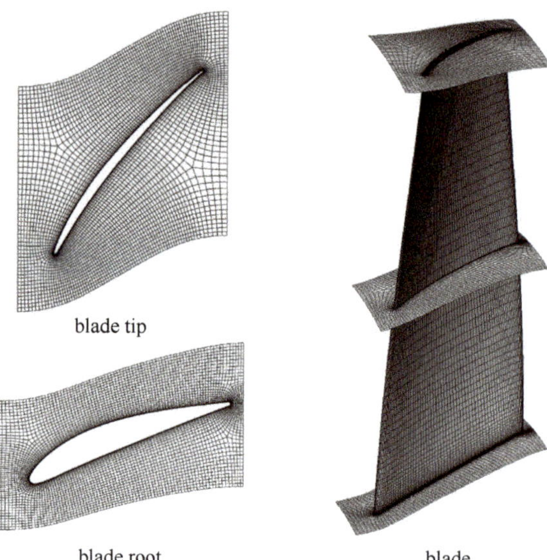

Figure 5. CFD mesh for the first-stage blade.

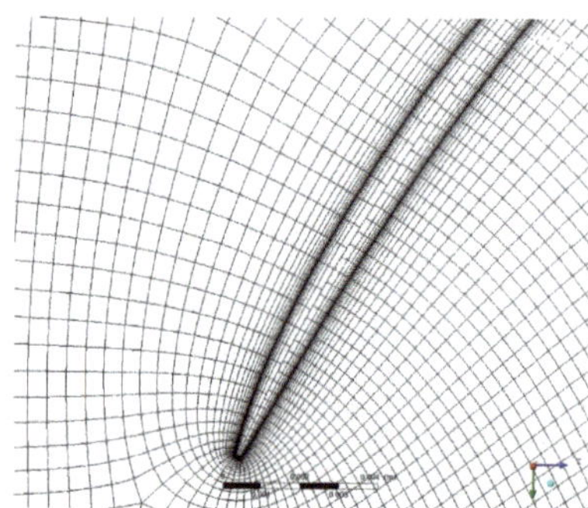

Figure 6. Mesh in the tip gap.

To reduce the required computing power, a single blade with cyclic symmetry along the lateral boundaries of the domain was modeled for each compressor stage. Stage interfaces (Mizing-Plane) between fixed and rotating domains were defined, which allowed for interpolation between interconnected grids, taking into account the laws of mass conservation.

The choice of the turbulence model depends on the nature of the turbulent flow, the required accuracy, the available time and computational resources. The SST K-ω turbulence model of Mentera was chosen as more accurate and reliable for the class of flows with a positive pressure gradient. The residual RMS error of 1×10^{-6} was assumed as a satisfactory condition of computational fluid dynamics (CFD) convergence and it was achieved after 600–870 iterations.

CFD results were used to estimate the compressor maps, as in another paper about this turboshaft [37]. The mass flow rate was measured at the outlet of the compressor. Each operating point corresponded to a certain mass flow, in the range from 4 to 11 kg/s.

2.4. Modeling Two-Phase Flow through IPS

The modeling of multiphase flows presents a number of difficulties compared to single-phase flows, since it is necessary to solve the equation of mass, amount of motion, and energy conservation for each phase separately. These equations are much more complicated than for the single-phase case, because they have additional terms that govern the exchange of mass and energy between phases. However, as a result of various concomitant physical phenomena and possible changes in the flow regime, the exact values of many parameters are not always known [6].

To study the operation of the inlet particle separator, we describe multiphase dispersion flows, in which there is a continuous and dispersion phase. The dispersed phases contained many particles distributed in a continuous phase. The Euler model and the ANSYS CFX software were used for modeling. The equations of mass, amount of motion, and energy conservation were solved separately for each phase. In the equations of motion, the interfacial drag force and other forces observed in multiphase dispersed systems were adopted. The calculations determined the local flow rate, temperature, and volume fraction of the dispersed phase. A granulation model was used to account for particle collision, friction, and density of the particles.

To simulate two-phase flow and obtain results on the velocity of motion and distribution of dust particles in the air, the following assumptions were made:

- Geometric model of the separator;
- Concentration and chemical composition of dust;
- Pressure and air velocity in the separator;
- Flow model: two-phase;
- Full pressure at the inlet to the engine: 101,325 Pa;
- Exit velocity: 150 m/s;
- Temperature: 288 K;
- Turbulence model: K-ϵ;
- Dust concentration at the inlet: 2 g/cm^2;
- Foreign particles material: quartz sand;
- Particle size: 10–50 µm in Case 1 and 50–100 µm in Case 2.

The IPS geometry, mesh, and other details are described in our previous publication [38]. Although particle separation is not the subject of this work, the selected calculation results are presented here to explain the wear process in engines operated with or without the IPS.

3. Results and Discussion

3.1. Chord Wear

The results of the dimensional control of the compressor blades show that the largest wear is observed at the blade tip, while a much smaller wear is exhibited near the root (Figure 7). The chord wear of Sections 2–4 can be expressed as a function of the wear of the tip section.

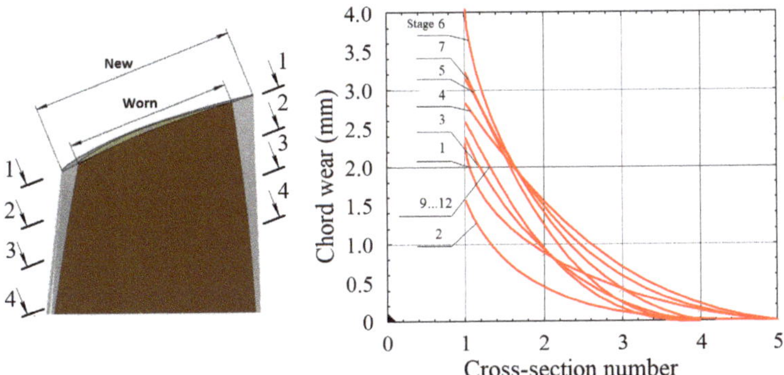

Figure 7. Chord wear of the compressor blade in specific cross-sections.

The statistical analysis carried out for individual stages revealed the wear patterns caused by the design features of the compressor [25]. Figure 8 shows the chord wear of the compressor blades for all stages. For engines operated without IPS, the largest wear is observed for the first-stage blades, while for engines operated with IPS, for the sixth-stage blades.

Engines belonging to the first category, operated without IPS, were grounded in accordance with the current standards for the maximum allowable chord wear of first-stage blades i.e., 2 mm at the tip. The effective TBO of the first category engines was only 150–200 FH, provided that it is nominally 1500 FH for this type of engine.

Engines from the second category, operated with IPS, were removed from service due to their operating parameters, such as n_1 speed and turbine inlet temperature beyond permissible limits. The effective TBO for the second category was 600–650 FH which indicates a partial effectiveness of the applied IPS [38].

Given the close correlation (R > 0.856) of chord wear among stages 2–12, it is possible to use these regression dependencies for modeling compressor wear. The chord wear of the sixth-stage blades is used as the independent variable. This choice is explained by the fact that, despite the obvious advantages of using the first stage for the dimensional control of blades, which is the most convenient in terms of monitoring and predicting the remaining useful life (RUL) without removing and disassembling the engine, it does not correlate well with the wear of the remaining stages (R < 0.4).

The chord wear of stages 2–12 correlates with the wear of the sixth stage and can be estimated using the linear regression.

$$c_i = a_i\, c_6 + b_i \tag{1}$$

The dependence of the chord wear on the compressor blades on the engine operating time in a dusty environment can be described by a second-order curve (Figure 9).

$$c_6 = \left(4.69 \times 10^{-6}\, t + 1.25 \times 10^{-3}\right) t\, \delta, \tag{2}$$

where δ is dust concentration in g/m^3. For the analyzed fleet, the average concentration $\delta = 1.6\,\text{g/m}^3$ can be assumed:

$$c_6 = 7.5 \times 10^{-6} t^2 + 0.02t. \tag{3}$$

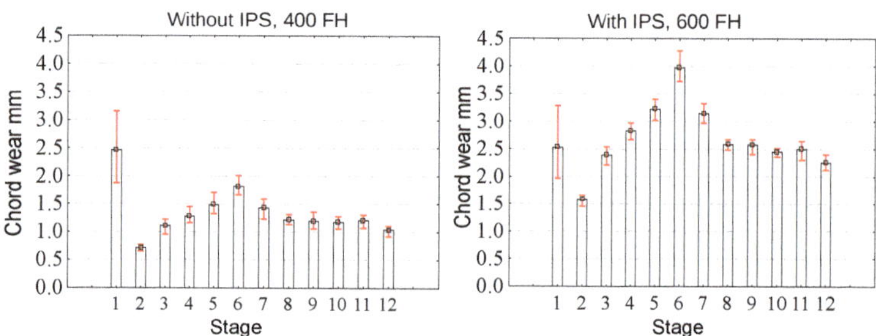

Figure 8. Average chord wear of compressor stages for an engine operated without inlet particle separator (IPS) or with IPS.

3.2. Modal Analysis of Blades

To assess the effect of wear on the natural frequency of the vibration, a modal analysis was performed (Figure 10). For each stage, an individual Campbell diagram was developed describing the nominal and worn profile of the aerofoil. Erosion significantly increases the natural frequencies of the blades. The analysis reveals several resonances excited by engine orders. Some stages were identified with a risk of resonance in the operating speed range (Figure 11).

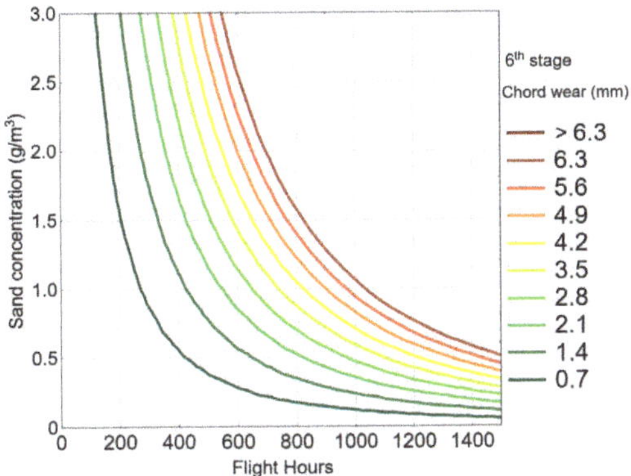

Figure 9. Chord wear of sixth-stage blades in function of flight hours.

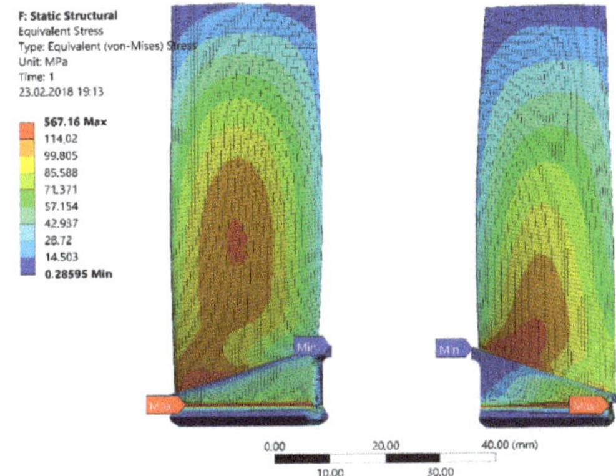

Figure 10. Modal and kinetostatic stresses of the first-stage blade with a nominal (**left**) and worn aerofoil (**right**) for $n_1 = 100\%$.

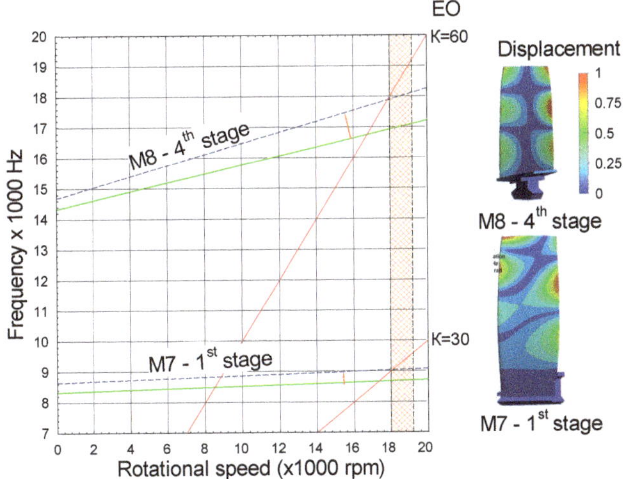

Figure 11. Impact of erosion on the Campbell diagram for critical modes and stages.

Stages and modes for which the risk of resonance is identified:

- M7 of the first stage resonates with the EO30 due to the number of the inlet guide vanes. This resonance is possible at 4 mm chord wear at the tip;
- M8 of the fourth stage resonates with the EO60 due to the number of the guide vanes of this stage. Resonance occurs when chord wear equals 6.3 mm.

3.3. Particle Separator

A particle separator cleans the air entering the engine from dust, sand, dry twigs, leaves, and other foreign objects during taxiing, take-off, and landing from unpaved runways or landing areas [39]. When IPS is turned on, with the engine running, hot air from the compressor enters the dust ejector nozzle. At the same time, as a result of centrifugal forces, part of the airflow entering the engine is pressed to the rear of the central fairing and enters the separator inlet. Most of the cleaned air passes through a separator to the engine inlet. Contaminated air, including foreign particles, enters the dust exhaust pipe, in which a vacuum is created by the ejector. Thus, particles are expelled into the atmosphere.

The results of modeling the speed and trajectory of foreign particles of various sizes flowing in the gas path showed that most of the particles are separated along the internal curved surface of the IPS due to the action of centrifugal forces. However, experience from the operation of helicopters in a dusty environment shows that the use of a particle separator of this type does not solve the problem of erosive wear of compressor aerofoils [38].

IPS is effective at separating large fractions of particles ranging from 50 to 100 µm, almost completely cleaning the air supplied to the engine compressor (Figure 12). If the particle size is relatively small—from 10 to 50 µm, about 20% of the particles enter the engine compressor bypassing the IPS separator under the influence of viscous forces following the flow. (Figure 13).

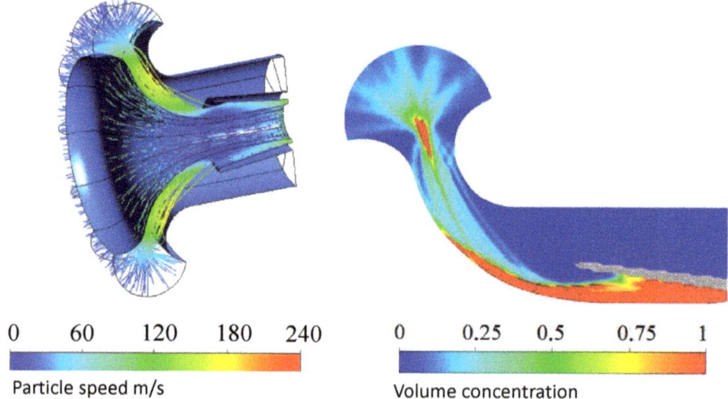

Figure 12. Velocity and concentration of large particles in the inlet of the turboshaft with IPS; particle size 50–100 µm.

3.4. Stall Margin Analysis

To assess the effect of erosion on the compressor performance, the flow for the nominal (initial) geometry of the blades, as well as the geometry corresponding to the engine operating time of 200, 400, 600, and 800 FH was simulated. Compressor pressure maps (Figure 14) and efficiency maps (Figure 15) were calculated and validated by comparison with available test data from a compressor rig and test cell.

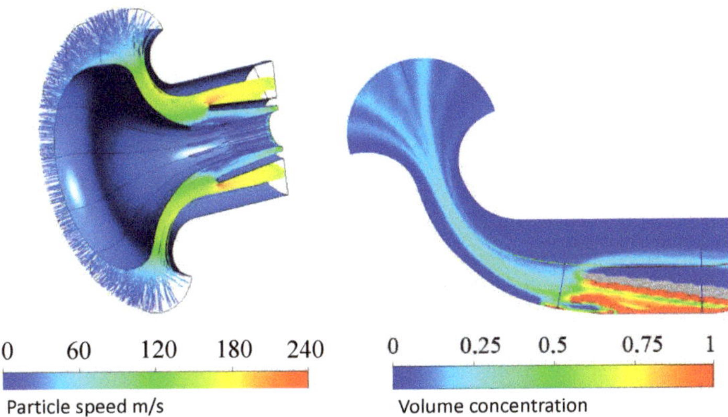

Figure 13. Velocity and concentration of small particles in the inlet of the turboshaft with IPS; particle size 10–50 μm.

The compressor maps plotted as a function of flight hours show that with increasing intensity of blade wear, the pressure ratio and its efficiency decreases. Consequently, the stall margin (SM) is also reduced (Figure 14). It was calculated using the following formula:

$$\mathrm{SM} = 100\% \left(\frac{\pi_{stall}/\dot{m}_{stall}}{\pi_{ss}/\dot{m}_{ss}} - 1 \right), \tag{4}$$

where π_{stall} is the pressure ratio for the stall line, and π_{ss} is the pressure ratio for the steady state line.

A decrease in the stall margin of the compressor by 15% causes the appearance of a surge during test-cell testing of TV3-117 engines. The analysis of flow through a compressor with blades of varying degrees of wear (Figures 16 and 17) showed that as a result of erosive wear of rotor stages 6–9, the compressor develops a stall in the high speed range, which leads to a surge. The reason for this phenomenon was chord reduction and an increase in the radial clearance. For this reason, the compressor operated in a dusty environment reaches its serviceability limit after 730–750 FH (Figure 18).

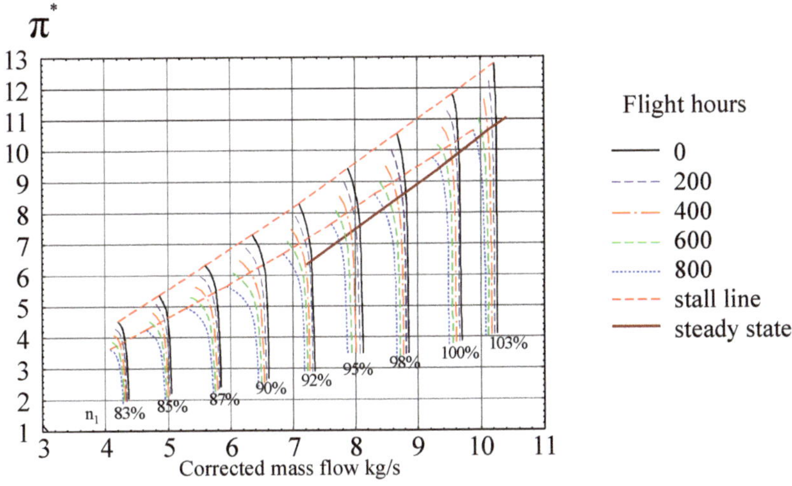

Figure 14. Dependence of pressure compressor maps on the operating time in a dusty environment.

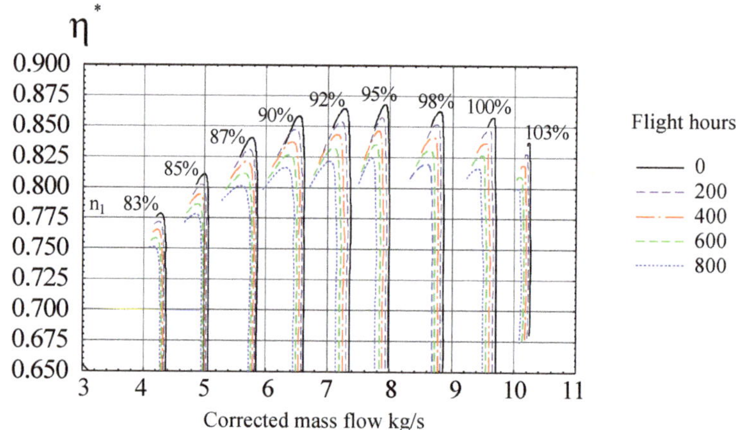

Figure 15. Compressor efficiency versus operating time in a dusty environment.

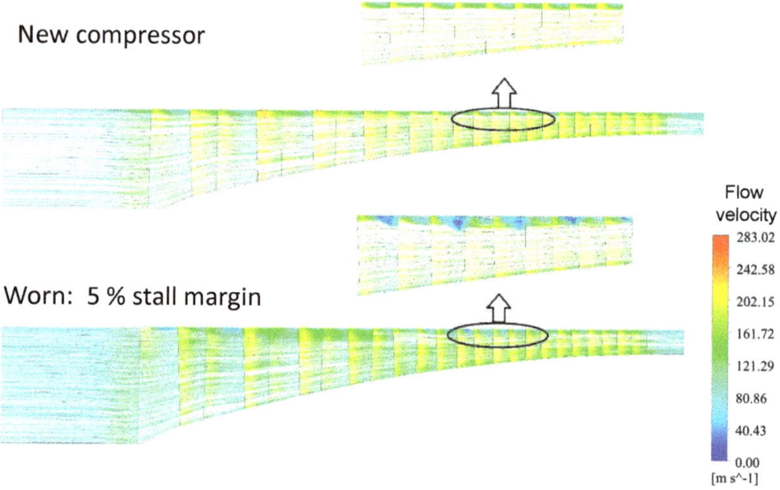

Figure 16. Flow velocity in the new and worn compressor.

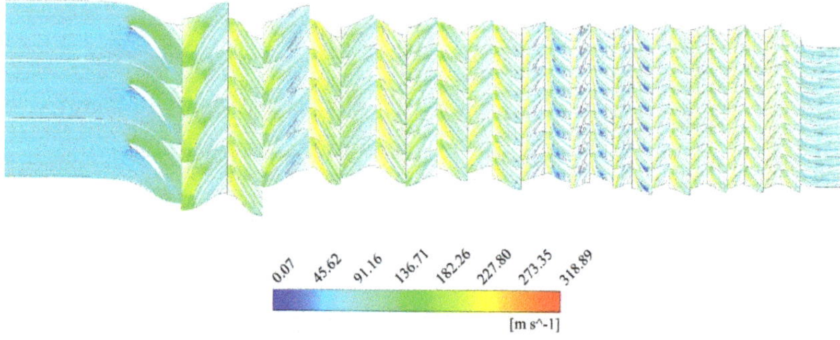

Figure 17. Flow velocity at 90% blade span.

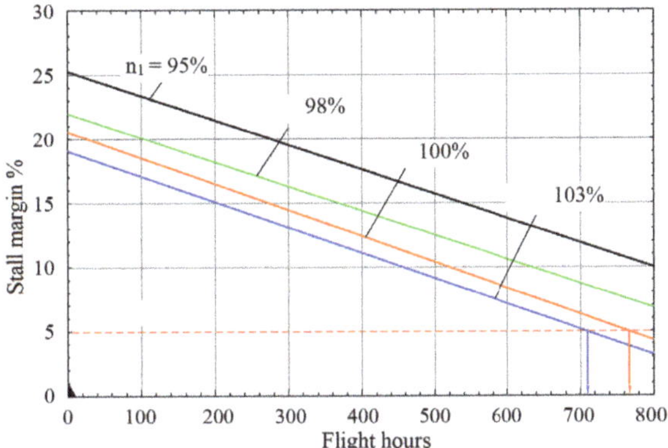

Figure 18. Stall margin of the compressor versus operating time in a dusty environment.

4. Conclusions

On the basis of the analysis of the geometry of the compressor blades of the TV3-117 engines operated in a dusty environment, it was found that the wear of the blades in all compressor stages occurs uniformly. When the particle separator is used, the largest chord wear is observed for stages 1–6. As a result of a close correlation of chord wear in stages 2–12, their wear can be related to the sixth stage.

An original methodology was developed to assess the influence of erosive wear of blades on compressor performance by modeling three-dimensional flow for various degrees of wear of all rotor stages. It involves measuring the chord wear of blades, calculating the natural frequencies of vibration whilst taking into account the aerofoil wear, numerical calculation of the compressor flow, and analyzing the onset of stall. This approach in combination with established patterns of chord wear over engine operating time allowed for the assessment of the serviceability limits of the blades.

On the basis of a modal analysis of compressor blades with different operating times, the dependency of blade vibration frequency on the chord wear for all stages was established. It was found that when the chord of the first-stage blades has worn more than 4 mm, M7 EO30 resonance can occur. Similarly, chord wear of the forth-stage blades higher than 5.1 mm causes M8 EO60 vibration. The HCF risk for the remaining stages is negligible.

The values of the maximum permissible chord wear of the blades of all stages are determined by the criterion of stall margin of the TV3-117 turboshaft. The chord wear of the sixth-stage blades of 6.19 mm is critical because it is accompanied by a decrease in the stall margin of the compressor by 15–17%, which indicates the appearance of a permanent stall at 770–790 FH. An additional slot for the optical inspection of the sixth stage was designed to allow for field maintenance (Figure 19).

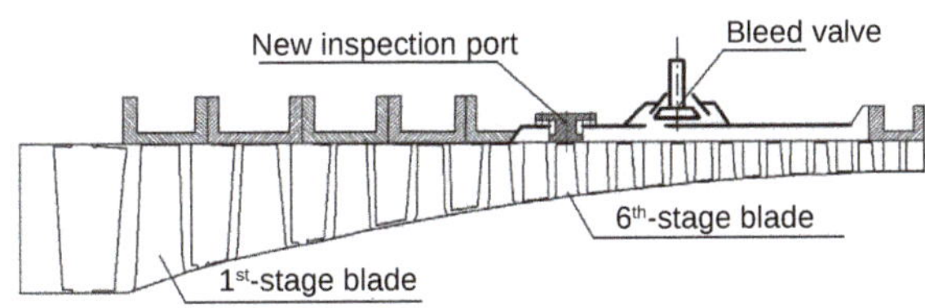

Figure 19. New slot designed to inspect the sixth stage.

The standards applicable today determine the maximum chord wear of a blade at the level of 2 mm at the tip, regardless of the use of particle separators. The criteria established in terms of natural vibration frequency and the stall margin allow for an increase in the time between overhauls (TBO) by 200 FH. The defined serviceability limits of the blades enable helicopter users to significantly reduce operating costs by extending the RUL of the engines operated in a desert environment.

The presented methodology can be utilized. However, further studies and more field data are needed to estimate the wear rate of compressor blades operated in other environments. For lower dust concentrations or more effective inlet protection, a more sophisticated model will be necessary to describe particle deposition in the compressor as well as the hot section. Large fleets may require machine learning tools to model significant amounts of diverse and variable data.

Author Contributions: Y.D. and D.P. conceived and designed the research; Y.D. processed and analyzed the data; Y.D. and D.P developed FEM and CFD models; D.P. and R.P. verified and evaluated the results. Y.D. and R.P. drew conclusions and produced the paper.

Funding: This research received no external funding.

Acknowledgments: This publication was prepared within the framework of the AERO-UA project, which has received funding from the European Union's Horizon 2020 research and innovation program under grant agreement No 724034.

Conflicts of Interest: The authors declare no conflict of interest. Motor Sich JSC had no role in the design, execution, interpretation, or writing the study. The views, information, or opinions expressed herein are solely those of the authors and do not necessarily represent the position of any organization.

Abbreviations

The following abbreviations and symbols are used in this manuscript:

δ	dust concentration
η^*	compressor efficiency
π^*	pressure ratio
n_1	Gas Generator Speed
CFD	Computational Fluid Dynamics
EO	Engine Order
FEM	Finite Element Method
FH	Flight Hours
FOD	Foreign Object Damage
HCF	High Cycle Fatigue
IPS	Inlet Particle Separator
ITWL	Air Force Institute of Technology in Warsaw
MDPI	Multidisciplinary Digital Publishing Institute
MRO	Maintenance, Repair, Overhaul
R	Pearson correlation coefficient
rpm	revolutions per minute
RUL	Remaining Useful Life
SM	Stall margin
TBO	Time Between Overhauls
TIT	Turbine Inlet Temperature
VIGV	Variable Inlet Guide Vanes

References

1. Hamed, A.A.; Tabakoff, W.; Rivir, R.B.; Das, K.; Arora, P. Turbine blade surface deterioration by erosion. *J. Turbomach.* **2005**. [CrossRef]
2. Szczepankowski, A.; Szymczak, J.; Przysowa, R. The Effect of a Dusty Environment upon Performance and Operating Parameters of Aircraft Gas Turbine Engines. In *STO-MP-AVT-272 Impact of Volcanic Ash Clouds on Military Operations*; The NATO Science and Technology Organization: Vilnius, Lithuania, 15–17 May 2017; pp. 1–13. [CrossRef]

3. Przysowa, R.; Gawron, B.; Kulaszka, A.; Placha-Hetman, K. Polish experience from the operation of helicopters under harsh conditions. *J. KONBIN* **2018**, *48*, 263–300. [CrossRef]

4. Doring, F.; Staudacher, S.; Koch, C.; Weißschuh, M. Modeling particle deposition effects in aircraft engine compressors. *J. Turbomach.* **2017**, *139*. [CrossRef]

5. Doring, F.; Staudacher, S.; Koch, C. Predicting the Temporal Progression of Aircraft Engine Compressor Performance Deterioration due to Particle Deposition. In Proceedings of the ASME Turbo Expo 2017: Turbomachinery Technical Conference and Exposition, Charlotte, NC, USA, 26–30 June 2017. [CrossRef]

6. Bojdo, N.; Filippone, A. A Simple Model to Assess the Role of Dust Composition and Size on Deposition in Rotorcraft Engines. *Aerospace* **2019**, *6*, 44. [CrossRef]

7. Abdullin, B.R.; Akmaletdinov, R.G.; Gumerov, X.S.; Nigmatullin, R.R. K issledovaniyu raboty GTD v zapylennoj atmosfere (Research of gas turbine engine operation in dust-filled atmosphere). *Vestnik Samarskogo Gosudarstvennogo Aehrokosmicheskogo Universiteta* **2014**, *5*, 95–102.

8. Kramchenkov, E. Issledovanie Ehrozionnogo Iznashivaniya Materialov (Study of Erosive Wear of Materials). Ph.D. Thesis, Gubkin Russian State University of Oil and Gas, Moscow, Russia, 1995.

9. Borkova, A.N. Erozionnaya Stojkost Aviacionnyx Materialov pri Soudarenii s Tverdymi (Pylevymi) Chasticami (Erosion Resistance of Aviation Materials in Collision with Solid (Dust) Particles). Ph.D. Thesis, All-Russian Institute Of Aviation Materials, Moscow, Russia, 2006.

10. *Gas Turbine Engine Environmental Particulate Foreign Object Damage [EP-FOD]*; The NATO Scientific and Technology Organization: Brussels, Belgium, 2019. [CrossRef]

11. Finnie, I.; Stevick, G.R.; Ridgely, J.R. The influence of impingement angle on the erosion of ductile metals by angular abrasive particles. *Wear* **1992**. [CrossRef]

12. Bitter, J.G. A study of erosion phenomena. Part II. *Wear* **1963**. [CrossRef]

13. Sheldon, G.L.; Kanhere, A. An investigation of impingement erosion using single particles. *Wear* **1972**. [CrossRef]

14. van der Walt, J.P.; Nurick, A. Erosion of Dust-Filtered Helicopter Turbine Engines Part I: Basic Theoretical Considerations. *J. Aircr.* **1995**, *32*, 106–111. [CrossRef]

15. Finnie, I. Some observations on the erosion of ductile metals. *Wear* **1972**. [CrossRef]

16. Khodak, M.O.; Vishnevskij, O.A. Eksperimentalni viprobuvannya ta prognozuvannya xarakteristik abrazivnoi znosostoijkosti materialiv aviacijnix GTD (Experimental tests and prediction of the abrasion resistance characteristics of materials of aviation gas-turbine engines). *Aviacionno-Kosmicheskaya Texnika i Texnologiya* **2006**, *7*, 114–123.

17. Evstifeev, A.; Kazarinov, N.; Petrov, Y.; Witek, L.; Bednarz, A. Experimental and theoretical analysis of solid particle erosion of a steel compressor blade based on incubation time concept. *Eng. Fail. Anal.* **2018**. [CrossRef]

18. Batcho, P.F.; Moller, J.C.; Padova, C.; Dunn, M.G. Interpretation of gas turbine response due to dust ingestion. *J. Eng. Gas Turbines Power* **1987**. [CrossRef]

19. Singh, D.; Hamed, A.L.; Tabakoff, W. Simulation of performance deterioration in eroded compressors. In Proceedings of the ASME 1996 International Gas Turbine and Aeroengine Congress and Exhibition, Birmingham, UK, 10–13 June 1996. [CrossRef]

20. Koch, C.C.; Smith, L.H. Loss sources and magnitudes in axial-flow compressors. *J. Eng. Gas Turbines Power* **1976**. [CrossRef]

21. Shpilev, K. *Jekspluatacija Letatelnyh Apparatov v Gorno-Pustynnoj Mestnosti [Operation of Aircraft in the Mountain-Desert Area]*; Voennoe izdatelstvo: Moscow, Russia, 1991; p. 224.

22. Tabakoff, W.; Hamed, A.; Wenglarz, R. Particulate flows, turbomachinery erosion and performance deterioration. *Von Karman Lect. Ser.* **1988**, *89*, 24–27.

23. Grigorev, V.; Zrelov, V.; Ignatkin, J.; Kuzmichev, V.; Ponomarev, B.; Shahmatov, E. *Vertoletnye Gazoturbinnye Dvigateli [Helicopter Gas Turbine Engines]*; Mashinostroenie: Moscow, Russia, 2007; p. 491.

24. *TV3-117VMA-SBM1V Series 4 and 4E Turboshaft Engine (Brochure)*; Motor Sich JSC: Zaporizhzhia, Ukraine, 2012.

25. Pavlenko, D.; Dvirnyk, Y. Zakonomernosti iznashivanija rabochih lopatok kompressora vertoletnyh dvigatelej, jekspluatirujushhihsja v uslovijah zapylennoj atmosfery (The laws of wear of the compressor rotor blades of the helicopter engines that are operated under the dust conditions). *Visnik Dvigunobuduvannja* **2016**, *1*, 42–51.

26. Vorobev, Y.; Romanenko, V. Analiz kolebanij lopatochnogo apparata kompressora GTD (Analysis of vibrations of gas turbine compressor blading). *Aviacionno-Kosmicheskaya Tehnika i Tehnologija* **2013**, *107*, 55–59.

27. Rzadkowski, R.; Gnesin, V.; Kolodyazhnaya, L.; Kubitz, L. Unsteady Forces Acting on the Rotor Blades in the Turbine Stage in 3D Viscous Flow in Nominal and Off-Design Regimes. *J. Vib. Eng. Technol.* **2014**, *2*, 3–9.

28. Witek, L. Crack propagation analysis of mechanically damaged compressor blades subjected to high cycle fatigue. *Eng. Fail. Anal.* **2011**. [CrossRef]

29. Witek, L. Crack Growth Simulation in the Compressor Blade Subjected to Vibration Using Boundary Element Method. *Key Eng. Mater.* **2014**, *598*, 261–268. [CrossRef]

30. Hamed, A.; Tabakoff, W. Experimental and numerical simulations of the effects of ingested particles in gas turbine engines. In Proceedings of the AGARD Conference Proceedings 558: Erosion, Corrosion and Foreign Object Damage Effects in Gas Turbines, Rotterdam, The Netherlands, 25–28 April 1994.

31. Itoga, H.; Tokaji, K.; Nakajima, M.; Ko, H.N. Effect of surface roughness on step-wise S-N characteristics in high strength steel. *Int. J. Fatigue* **2003**. [CrossRef]

32. Murakami, Y.; Tsutsumi, K.; Fujishima, M. Quantitative evaluation of effect of surface roughness on fatigue strength. *Nippon Kikai Gakkai Ronbunshu* **1996**. [CrossRef]

33. Dyblenko, Y.M.; Selivanov, K.S.; Valiev, R.R.; Skryabin, I.V. Issledovanie gazoabrazivnogo iznosa obrazcov iz titanovogo splava VT-6 s nanostrukturirovannymi zashhitnymi pokrytiyami (Investigation of gas-abrasive wear of VT-6 titanium alloy samples with nanostructured protective coatings). *Vestn. Ufim. Gos. Aviac. Texnicheskogo Univ.* **2011**, *15*, 83–86.

34. Ivchenko, D.; Shtanko, P. Ob ustalostnom mexanizme gazoabrazivnoj erozii detalej gazovozdushnogo trakta vertoletnyx GTD [On the fatigue mechanism of gas-abrasive erosion of parts of the gas-air duct of a helicopter gas turbine engine]. *Visnik Dvigunobuduvannya* **2009**, *2*, 12–14.

35. Batailly, A.; Legrand, M.; Cartraud, P.; Pierre, C. Assessment of reduced models for the detection of modal interaction through rotor stator contacts. *J. Sound Vib.* **2010**, *329*, 5546–5562. [CrossRef]

36. Ma, H.; Wang, D.; Tai, X.; Wen, B. Vibration response analysis of blade-disk dovetail structure under blade tip rubbing condition. *J. Vib. Control* **2017**, *23*, 252–271. [CrossRef]

37. Marakueva, O. Aerodynamic design and optimization of blade configuration in an inlet stage of an aircraft engine compressor. In Proceedings of the 29th Congress of the International Council of the Aeronautical Sciences, St. Petersburg, Russia, 7–12 September 2014.

38. Dvirnyk, Y.; Pavlenko, D. Zakonomernosti techenija dvuhfaznogo potoka vo vhodnom ustrojstve vertoletnyh GTD (Laws of multiphase flow behavior in the inlet of a helicopter engine). *Aviacionno-Kosmicheskaya Tehnika i Tehnologija* **2017**, *7*, 30–37.

39. Filippone, A.; Bojdo, N. Turboshaft engine air particle separation. *Prog. Aerosp. Sci.* **2010**, *46*, 224–245. [CrossRef]

Article

Predesign Considerations for the DC Link Voltage Level of the CENTRELINE Fuselage Fan Drive Unit

Stefan Biser [1,2,*,†], **Guido Wortmann** [1,†], **Swen Ruppert** [3,†], **Mykhaylo Filipenko** [1,†], **Mathias Noe** [2] **and Martin Boll** [1,†]

1   Rolls-Royce Limited & Co. KG, Willy-Messerschmitt-Str. 1, 82024 Taufkirchen, Germany;
    guido.wortmann@rolls-royce.com (G.W.); mykhaylo.filipenko@rolls-royce.com (M.F.);
    martin.boll@rolls-royce.com (M.B.)
2   Karlsruhe Institute of Technology—Institute of Technical Physics, Hermann-von-Helmholtz-Platz 1,
    76344 Leopoldshafen-Eggenstein, Germany; mathias.noe@kit.edu
3   Siemens AG, Anton-Bruckner-Str. 2, 91052 Erlangen, Germany; swenruppert@t-online.de
*   Correspondence: stefan.biser@rolls-royce.com
†   These authors contributed equally to this work.

Received: 2 October 2019; Accepted: 14 November 2019; Published: 20 November 2019

Abstract: Electric propulsion (EP) systems offer considerably more degrees of freedom (DOFs) within the design process of aircraft compared to conventional aircraft engines. This requires large, computationally expensive design space explorations (DSE) with coupled models of the single components to incorporate interdependencies during optimization. The purpose of this paper is to exemplarily study these interdependencies of system key performance parameters (KPIs), e.g., system mass and efficiency, for a varying DC link voltage level of the power transmission system considering the example of the propulsion system of the CENTRELINE project, including an electric motor, a DC/AC inverter, and the DC power transmission cables. Each component is described by a physically derived, analytical model linking specific subdomains, e.g., electromagnetics, structural mechanics and thermal analysis, which are used for a coupled system model. This approach strongly enhances model accuracy and simultaneously keeps the computational effort at a low level. The results of the DSE reveal that the system KPIs improve for higher DC link voltage despite slightly inferior performance of motor and inverter as the mass of the DC power transmission cable has a major share for a an aircraft of the size as in the CENTRELINE project. Modeling of further components and implementation of optimization strategies will be part of future work.

Keywords: electric propulsion; aircraft; CENTRELINE; DC link voltage level; analytical model; design space exploration

1. Introduction

While noise levels have been reduced drastically, absolute emissions from aviation, especially carbon dioxide (CO_2) and nitrogen oxide (NO_x), have been continuously on the rise throughout the last decades, accounting for 2–3% of global greenhouse gas emissions. The large improvements in fuel efficiency are thwarted by the steady increase in passenger numbers [1,2]. The National Aeronautics and Space Administration (NASA) N+3 fundamental technology research for subsonic fixed-wing aircraft [3,4], the Civil Aviation Organization (ICAO) [5,6] as well as the Aviation Transport Action Group (ATAG) [2] and the European Commission's Flightpath 2050 [7] set aggressive target levels for greenhouse gas emissions in the future—up to 75% less fuel burn and 90% NO_x emissions by 2050.

To reach those ambitious objectives, a significant improvement in terms of fuel consumption is necessary. These objects can be seen as one of the main drivers for the development of novel propulsion systems [8], resulting in a large increase in interest on partly or fully electrified aircraft [9,10].

The change to electric-based propulsion systems not only affects current designs, it completely changes the setup. Three types of EP for aircraft can be distinguished by their level of hybridization [9]. Turbo-electric propulsion systems have a total hybridization of power, i.e., they generate all power from burning fuel in gas turbines [11,12]. Hybrid-electric concepts involve power generation from burning fuel and electric storage [8,13]. All-electric propulsion systems only have batteries on-board for energy storage [14].

An often-discussed point is the choice of the DC link voltage level for power transmission. The voltage level of state-of-the-art aircraft is currently limited to ±270 V DC to avoid corona at typical flight altitudes [15,16]. Increasing electric power demand of aircraft requires higher voltage levels to reduce weight and keep transmission efficiency at a high level [14,16,17]. Considering DC link voltages from 1000 V to over 5000 V, several studies showed that the mass of the power transmission cable resulting from the choice of the DC link voltage level is an important driver in system mass [15,17]. Up to now, the sensitivity of the electric machine and inverter design for electric aircraft on the DC link voltage has not yet been investigated in a coupled research.

In this paper, we elaborate this problem, having a closer look on the propulsion system of the CENTRELINE project [18] where 8 MW of propulsion power are required. In contrast to conventional aircraft powered by gas turbines, especially turbo-electric and hybrid-electric propulsion systems have significantly increased complexity on the system level and mass. Due to the increased number of components with a large number of parameters influencing the mission performance, a large design space is opened up [9,12]. Brelje and Martins [9] reviewed many approaches which try to incorporate sizing methods for EP on different levels of detail into the conventional aircraft sizing procedure. However, most of them use simple correlations based on stochastic regression or fixed parameters for specific power and efficiency of the components within the electric drivetrain [8,12,19–23]. Accordingly, EP is still poorly understood due to the low fidelity models that are used [9]. Thus, they proposed to develop a fully coupled, multidisciplinary analysis and optimization (MDAO) for electric aircraft.

As such models require enormous calculation power and long calculation time, we follow a different approach in this paper where each component is described by a physically derived, analytical model linking several subdomain models specific to the component together; e.g., for an electric machine, these are electromagnetics, structural mechanics and thermal analysis. All such components models are finally set up as coupled model for the complete system. This approach aims to strongly enhance the prediction accuracy, especially in terms of plausibility, reliability, and traceability of the results, while at the same time keeping the computational effort low to perform large design space explorations.

The paper is structured as follows: Firstly, the CENTRELINE project and the reference propulsion architecture are shortly introduced. Secondly, the specific challenge is presented, and the sizing process of the components is explained in more detail. The results of the large design space exploration are discussed component-wise as well as for the combined system. After a discussion of the results, an outlook of future work will be given.

2. CENTRELINE Project Overview

CENTRELINE is a project funded by the European Union (EU), serving as a demonstrator for propulsion-aircraft integration with the goal of validating results of former studies on boundary layer ingestion (BLI) such as the FP7 project DisPURSAL [24]. The major change in aircraft design is the implementation of the so-called fuselage fan (FF) drive unit which will be installed in the fuselage aft-end [18]. This device ingests and re-energizes the boundary layer of the fuselage to compensate viscous drag in the fuselage boundary layer flow [18,25].

While DisPURSAL [24] focussed on a conventionally driven fuselage propulsor by a third gas turbine, CENTRELINE will evaluate the possible advantages of a fully turbo-electric drivetrain arrangement to overcome shortcomings of a gas turbine in the aft section such as flow losses associated with the air intake duct, vibration issues and high maintenance costs [26]. The electric powertrain consists of two power generation sets, an electric power transmission system as well as a single fuselage propulsor located in the rear of the aircraft, as shown in Figure 1. Each generation is composed of a wing-podded, advanced geared turbofan (GTF), an electric generator that converts mechanical to electric power, and an AC/DC rectifier. The electric generator is mechanically coupled to an additional free power turbine located in the low-pressure turbine (LPT) section of the GTF even though this is in the hot section, as this has the minimal influence on engine stability. The rectifier will be placed in the nacelle of the GTF to evaluate the potential of air-cooled power electronics [23].

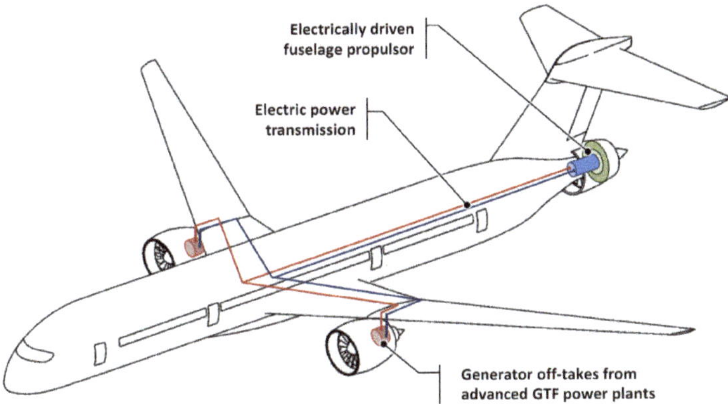

Figure 1. H2020 CENTRELINE architecture (turbo-electrically driven fuselage wake-filling) [18].

A DC power transmission system is chosen, consisting of eight parallel lanes to transfer the power as well as protection and switching devices to control it. In the rear, a single propulsion unit including eight parallel DC/AC inverters, one electric motor and the fuselage fan itself is mounted. The inverter converts electric DC to AC power, while the motor converts it to mechanical power to drive the fuselage fan directly without any additional gearbox [27].

Several limitations with regards to installation space of inverter and electric motor (depicted in Figure 2), derived by aerodynamic calculations, must be considered during the sizing process.

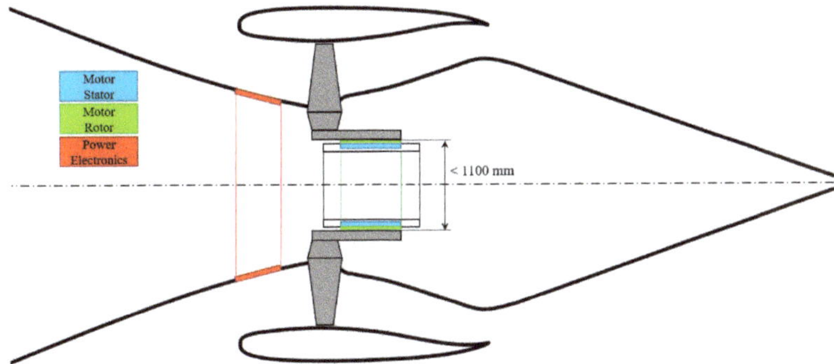

Figure 2. Fuselage fan drive unit integration with installation space limitations (taken from [27] and modified).

3. Methodolgy

3.1. Problem Setup

Brelje and Martins [9] as well as Gesell et al. [12] stated that EP is not applicable as drop-in replacement of existing aircraft propulsion unit but has to be optimized for aircraft and mission requirements. As outlined in Section 1, the proposed approach of sizing components of the electric drivetrain with physically derived, analytical models systematically improves the level of model details in order to raise awareness of importance of a better understanding of EP systems for aircraft. The average run-time per system design point is within the small digits of seconds on an Intel® Core™ i7-6600U CPU @ 2.60 GHz.

The approach was applied to the FF drive unit of the concept aircraft, considering only a reduced system with the electric motor, the DC/AC inverter and the cabling, while other components are neglected. This study aimed to investigate the influence of a change in DC link voltage level of component as well as system KPIs, especially mass and efficiency. Figure 3 shows the block diagram of the setup.

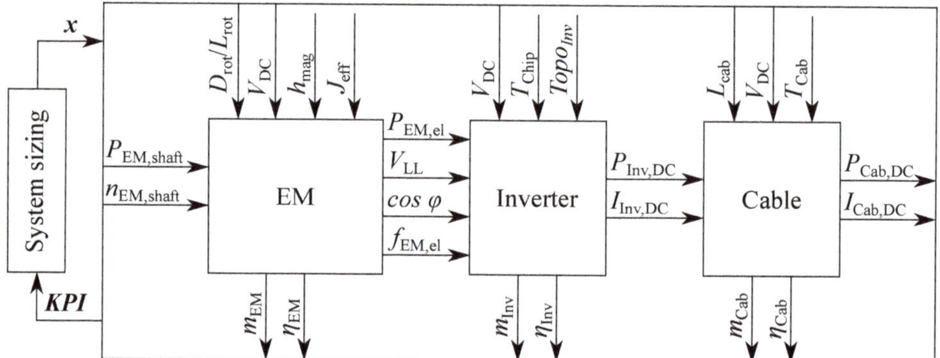

Figure 3. Block diagram of FF drive unit including motor, inverter and cable. For sake of simplicity only one inverter-cable configuration instead of all eight parallel lanes which feed the winding systems of the electric motor is shown.

The diagram only shows the parameters which are relevant for the scope of this study, in detail:

- general requirements (shaft power $P_{EM,shaft}$, rotational speed $n_{EM,shaft}$ and diameter-to-length-ratio D_{rot}/L_{rot} of the electric machine as well as the cable length L_{Cab}) which must be fulfilled by the component sizing process;
- global variation parameters which are relevant for more than one component (DC link voltage level V_{DC});;
- local variation parameters which only influence the design of one component (conductor current density J_{eff}, inverter topology $Topo_{Inv}$, allowable chip T_{chip} and cable temperature T_{Cab});
- output/input data, i.e., results from one component sizing process which are necessary for the design process of another component (line-to-line-voltage at the motor terminal V_{LL}, power factor $cos\ \varphi$, electric frequency $f_{EM,el}$ and the input DC current of the inverter $I_{Inv,DC}$); and
- key performance indicators (KPI), representing relevant information, such as mass m and efficiency η of the components.

General requirements as well as global and local variation parameter define a unique input vector x, which results in a unique output vector **KPI**. The necessity of a fully coupled system sizing model is elucidated using the example of the DC link voltage level. Depending on it, the winding layout as well as insulation aspects in the motor vary, which in term influences the selection of semiconductor models during the inverter sizing process. Furthermore, the selection of the DC link voltage level has significant impact on the sizing of the cabling as it is directly proportional to the current. Those effects can only be made visible if the interdependencies between components are considered correctly.

3.2. Requirements, Constraints and Design Target

The requirements for the electric motor as well as the semiconductor technology of the inverter and the length of the transmission cable stem from the aircraft design process. The electric motor has eight parallel winding systems each fed by a DC/AC inverter and connected to a DC transmission cable, transferring an eighth of the total power, respectively. The range of the DC link voltage of the transmission system is based on a pre-assessment of available semiconductor blocking voltage levels [27]. As the semiconductor module selection is based on the chosen inverter topology and the DC link voltage, five different topologies including two-, three-, and five-level topologies, are considered. Additionally, several internal parameters for each component are varied within reasonable ranges.

Table 1 summarizes the most important global design requirements as well as global and local variation parameters.

Table 1. Global requirements and input ranges for global and local variation parameters.

Component	Parameter	Range/Value	Unit	Description
Electric machine	$P_{EM,shaft}$	8000	kW	mechanical shaft power
	$n_{EM,shaft}$	2100	min^{-1}	shaft rotational speed
	τ_{avg}	6–8	N cm^{-2}	specific thrust
	D_{rot}/L_{rot}	1–2	-	diameter-length-ratio rotor
	p	12–50	-	number of polepairs
	Q	36–141	-	number of stator slots
	h_{mag}	9–12	mm	thickness of magnets
	J_{eff}	17.5	A mm^{-2}	conductor current density
DC/AC Inverter	$P_{EM,el}$	1000	kV A	electric output power
	n_{Inv}	8	-	number of parallel inverters
	$Topo_{Inv}$	2L, 3LNPC, 3LTNPC, 3LFC, 5LSMC	-	topology
	$Tech_{Inv}$	SiC	-	semiconductor technology
	$T_{cool,in}$	75	°C	cooling inlet temperature
	T_{chip}	150	°C	allowable chip temperature
	$r_{th,ca}$	105	K mm^2 W^{-1}	area specific thermal resistance case to cooling fluid
DC transmission cable	$P_{Inv,DC}$	1000	kW	DC power per transmission line
	L_{Cab}	86.0	m	total length
	V_{DC}	1500–3000	V	DC link voltage
	T_{cond}	120	°C	allow. conductor temperature
	T_{amb}	55	°C	ambient temperature

Besides, several constraints, especially concerning the available installation space for motor and inverter, must be considered, as shown in Figure 2. Based on the given input, a large design space exploration is performed. The design target is to identify Pareto-optimal design points with respect to the specified KPIs, i.e., system mass m_{sys} and efficiency η_{sys}, for the given input sets x according to Fang and Qin [28], such that

$$\min_{x}(\{m_{sys}(x), (1 - \eta_{sys}(x))\}) \tag{1}$$

where $m_{sys} = \sum_i m_i$ and $\eta_{sys} = \prod_i \eta_i$. Pareto-optimality is chosen as criterion to enable an evaluation of designs with concurring KPIs.

Within the next section, the sizing process of each component is explained in more detail.

3.3. Detailed Description of Sizing Process of Components

3.3.1. Electric Machine

Several machine types have been studied extensively for electric aircraft [29]. Surface-mounted, permanent magnet synchronous machines (PMSM) have been identified as the most promising technology. PMSM have several DOFs, e.g., the number of poles and slots, the choice of materials for magnets and the laminated core of rotor and stator, the diameter-length-ratio of the rotor D_{rot}/L_{rot} or the stator electric current density J_{eff}, which represent the local variation parameters.

The iterative design process of the electric machine includes five disciplines and starts with a defined set of requirements that the final design must fulfil. In the first step, the geometrical layout of the machine is determined. Afterwards, the electromagnetic design includes sizing the magnetic circuit under no-load and load conditions with a non-linear lumped parameter model (e.g., [30,31]), adjusting phase current to reach the required shaft torque $T_{EM,shaft}$ as well as the configuration of the winding layout. The electromagnetic calculation is done with Simcenter SPEED 13.06 [31]. The structural mechanics section covers the sizing of the retention sleeve for the magnets for different load cases as well as designing rotor shaft and stator housing. The thickness calculation is based on an analytical press-fit model which evaluates the tangential stresses with respect to the radial overclosure of the fit [32,33]. A further important part of machine design is insulation coordination based on IEC 60664 [34] to ensure secure operating voltage levels. To conclude the design process, a thermal analysis of the motor is performed to ensure that waste heat can be dissipated, which especially dimensions the cooling channels. The allowable conductor current density depends on the heat transfer into the cooling medium, which in turn is a function of the thickness of the insulation hindering heat flux and thus decreases allowable conductor current density. Considering the DC link voltage during the motor design thus facilitates a more realistic design.

Passive masses of the machine including rotor shaft, bearings, housing, and terminal box are hard to calculate as they strongly depend on the actual design as well as installation issues. According to a benchmark of several lightweight motors [35], a factor of two was used to account for the passive masses. This is a fair assumption as the power and rotational speed of the machine is fixed and thus the dimensioning torque for all designs identical.

After those design steps, the result is checked against the requirements. If they are not fulfilled the design process enters a further loop, otherwise the design is valid and the specifications are extracted, e.g., to serve as input for a further component such as the DC/AC inverter.

3.3.2. DC/AC Inverter

DC/AC inverter designs can be realized in different topologies, differing in the height of the single-step output voltage and the total harmonic distortion (THD) [27,36,37]. This study only considers two-level (2L-), three-level (T-type) neutral point clamped (3L(T)NCP-) and fly-cap (3LFC-), and five-level stacked multicell (5LSMC-) DC/AC voltage source inverters. The topologies are explained in detail by the work of Krug [36,37], on which most parts of the design process are based on.

First, the voltage configuration of switches and diodes for the given operation point are determined [36]. A scalable semiconductor characteristic is established from a database with commercially available semiconductor modules and their properties which is then used to perform a coupled calculation of electric and thermal performance. Based on the operation point, the losses of the semiconductor modules are calculated analytically [36,38,39]. The semiconductor modules as well as the heat sink, i.e., a cooling plate, are sized such that the chip temperature increase due to the losses at the calculated operating point is below a certain limit and that the waste heat can be removed via the cooling plate at steady-state conditions. As silicon-carbide MOSFETs (SiC) chips are used, a maximum chip temperature of 150 °C is assumed.

Furthermore, the DC link capacity is calculated based on the expected current ripple calculated analytically using a method provided by Krug [36]. Again, a database of commercially available capacitors is used. The sizing process also includes a mass estimation of the gate driver units and a simple housing but misses a calculation of electromagnetic interference (EMI) filters on AC and DC side. To account for the additional mass of the filters, a factor of two is applied.

3.3.3. DC Power Transmission Cable

The power transmission cables are routed from the generators through pylon and wing to the fuselage center to the fuselage aft, where the motor is located. The total length of the cables adds up to 86.0 m. Due to the long distance, aluminum cables are chosen instead of copper cables because of their higher ratio of electrical conductivity to specific weight. A transmission system with eight parallel DC lanes is used to mitigate the risk of a complete power loss in the event of a cable failure [27].

The geometrical design of the DC transmission cable only consists of a circular conductor material, i.e., copper or aluminum, and a concentric layer of insulation material, as electromagnetic field effects and thus shielding can be neglected. The cable sizing process is based on a coupled model of electrical and thermal analysis as well as insulation coordination. The electrical domain covers the calculation of electrical parameters such as voltage levels, currents and resistances as well as the ohmic losses in the conductor considering the conductor material. Insulation coordination determines the necessary thickness of insulation to avoid arcing according to IEC 60664 [34]. A thermal analysis determines the size of the conductor by iteratively solving a one-dimensional, cylindrically symmetrical heat equation [40]. The size of the conductor is chosen such that the losses can be dissipated via natural convection and radiation at given ambient temperature T_{amb}, and the maximum admissible conductor temperature T_{cond} at the insulation interface is not exceeded.

4. Results and Discussion of Design Space Exploration

The results of the design space exploration are discussed for the single components as well as for the combined system.

4.1. Results for Components

4.1.1. Electric Machine

First, the results of the design space exploration for the electric motor are evaluated. Figure 4 shows the gross mass m_{EM} (active and passive masses) of the electric machine against its electric efficiency η_{EM} at full load for three different DC link voltages V_{DC}, ranging from 1500 V to 3000 V. Each marker represents a valid motor design according to the design process described in Section 3.3.1; however, the graph only shows a fraction of the calculated design for better readability.

The Pareto fronts for each voltage level with respect to minimum mass and maximum efficiency (i.e., minimum losses) are depicted as line plots. As can be seen, the higher is the voltage level, the heavier are the machines for identical efficiency levels. This is a result of the increasing thickness of slot, phase, and turn insulation due to the increased voltage level, which in turn leads to a larger necessary slot area. This effect is intensified by the larger cooling channels required due to inferior heat transfer through the thicker insulation and not compensated by lower current levels.

The influence of increased DC link voltage level V_{DC} on the machine efficiency is minor, as especially the copper losses depend on the electric current density of the wire which was unchanged in this study. The slight decrease of efficiency is due to the increased core losses as the absolute loss is proportional to the iron mass at roughly constant specific iron losses. The best machines reach power densities up to $11.0 \, \mathrm{kW \, kg^{-1}}$, torque densities up to $50.0 \, \mathrm{N \, m \, kg^{-1}}$ and efficiencies up to 98.0%. The peripheral speed of the rotor is moderately high ($v_{rot} \approx 90 \, \mathrm{m \, s^{-1}}$) due to the large size ($D_{rot} \approx 0.4 \, \mathrm{m}$), wherefore the high power density can be attributed mainly to a high electromagnetic and thermal utilization of materials. This also reflects in a high current density in the stator ($J_{eff} = 17.5 \, \mathrm{A \, mm^{-2}}$), a value which is challenging but not impossible with state-of-the-art cooling techniques [41].

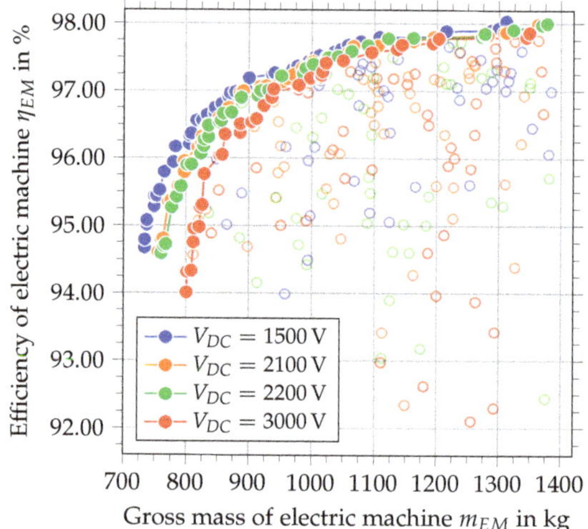

Figure 4. Gross mass against efficiency of electric machine for different DC link voltage levels V_{DC} ($P_{EM,shaft}$ = 8.0 MW, $n_{EM,shaft}$ = 2100 min^{-1}, J_{eff} = 17.5 A mm^{-2} and $f_{EM,el}$ = 420–1750 Hz). Line plots indicate Pareto-fronts for different voltage levels. Not all results shown for better readability (≈5%).

4.1.2. DC/AC Inverter

As the design of the inverter is based on the use of databases of commercially available semiconductor modules with discrete blocking voltages and capacitors and utilizes different topologies, a different visualization method is chosen. Masses and efficiencies for different topologies are plotted against varying DC link voltage.

Figure 5 shows mass respectively efficiency of several inverter topologies against varying DC link voltage. In Figure 5a, the discrete steps at certain voltages can be seen clearly, which are determined by the allowable utilization accounting for derating at high flight altitudes [27] and the blocking voltage of the semiconductors as well as the number of inverter levels. The maximum efficiency within the different sections which can be reached increases with higher DC link voltages V_{DC} for the different topologies except for the 2L inverter due to its inferior loss characteristic at high voltages.

Within the sections, the mass of the inverters decreases slightly due to the lower currents as well as the decreasing current ripple on DC link capacitors and the fly caps. The volatility in mass results from the discrete choice of capacitors out of a database. The absolute higher mass for further increased DC link voltage can be explained by the specifically higher mass of the semiconductor modules with higher blocking voltages (see Figure 5b).

The most lightweight inverters reach power densities up to $P_{Inv,AC}/m_{Inv}$ ≈ 55 kV A kg^{-1} (gravimetric) and $P_{Inv,AC}/V_{Inv}$ ≈ 50 kV A dm^{-3} (volumetric) at efficiencies of up to η_{Inv} = 99.0% for maximum chip temperature T_{chip} = 150 °C at a fluid inlet temperature T_{amb} = 75 °C for a specific resistance $R_{th,ca}$ = 0.012 K W^{-1} of the heat sink surface to ambient (in this case the liquid). Those values are in good accordance with a literature review [42].

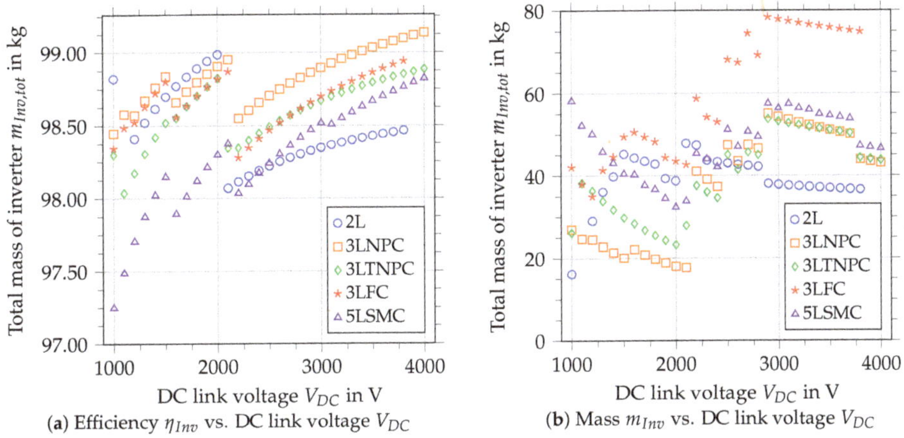

(a) Efficiency η_{Inv} vs. DC link voltage V_{DC}

(b) Mass m_{Inv} vs. DC link voltage V_{DC}

Figure 5. Efficiency η_{Inv} and gross mass m_{Inv} for varying DC link voltage V_{DC} and different inverter topologies ($P_{EM,el} = 1.0\,\text{MVA}$, $\cos\varphi = 0.85$, $T_{chip} = 150\,^\circ\text{C}$, $T_{amb} = 75\,^\circ\text{C}$ and $f_{EM,el} = 1000\,\text{Hz}$).

4.1.3. DC Transmission Cable

Figure 6 shows the DC link voltage vs. the cable mass for the eight DC lane configuration specified in Section 3.2 for aluminum as well as copper conductors. While aluminum has only 60% of the electrical conductivity compared to copper, its density is roughly a third, resulting in 50% of the cable mass per ampere. The mass of the cable reduces significantly with increasing voltage as the current decreases and with that the Ohmic losses, which are the sizing parameter for the conductor cross-section.

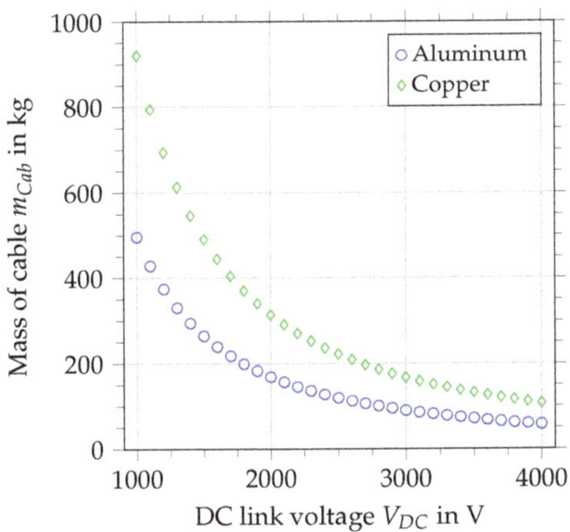

Figure 6. Cable mass m_{Cab} vs. varying DC link voltage V_{DC} of the DC power transmission cable ($P_{Inv,DC} = 1.0\,\text{MW}$, $T_{Cab} = 120\,^\circ\text{C}$ and $L_{Cab} = 86.0\,\text{m}$).

With increasing DC link voltage level V_{DC}, the cable mass decreases roughly indirectly. For unscreened, stranded aluminum cables with silicone rubber insulation comparable to those in [43], the absolute mass of the insulation only increases slightly for higher voltage levels as the current and thus the conductor cross-section decreases significantly. The relative mass share of the insulation compared to the total cable mass rises from roughly 15% for 1500 V to 35% for 3000 V.

A high DC link voltage is thus favorable, while a point of minimum cable mass might exist for even higher voltages when insulation starts to dominate the total cable mass.

4.2. System Results

Figure 7 shows the results of the CENTRELINE DSE study. Figure 7a outlines the results with respect to the total system mass m_{Sys} and system efficiency η_{Sys} of the FF drive unit including the electric motor as well as eight DC/AC inverters and DC power transmission lanes.

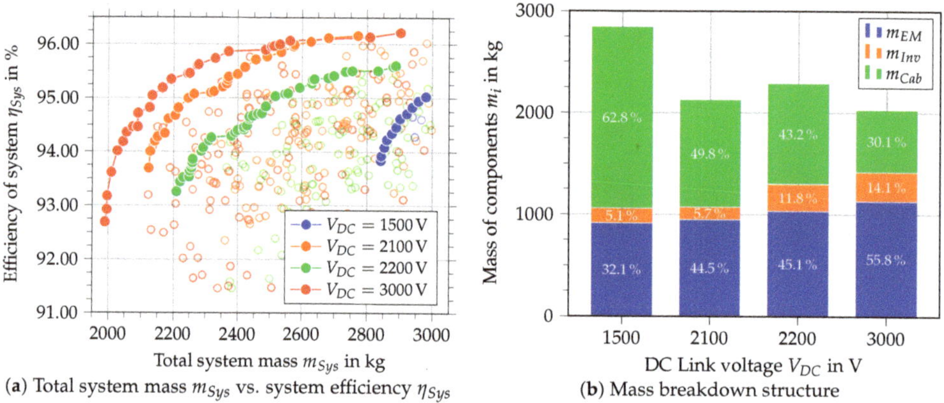

(**a**) Total system mass m_{Sys} vs. system efficiency η_{Sys} (**b**) Mass breakdown structure

Figure 7. CENTRELINE DSE study results: (**a**) Total system mass m_{Sys} against system efficiency η_{Sys} of FF drive unit for different DC link voltage levels V_{DC}. Line plots indicate Pareto-front for different voltage levels. Not all results shown for better readability ($\approx 5\%$). (**b**) Breakdown of component masses m_i including electric machine, inverters, and cables for Pareto-optimal point for different DC link voltages V_{DC} at similar system efficiency $\eta_{Sys} \approx 94.0\%$.

Two effects superpose each other. First, the general trend implies that, with higher DC link voltage level V_{DC}, lower system masses and higher system efficiencies are possible. As both motor and inverter tend to be heavier the mass of the cable must have a major share and its decrease in mass overcompensates the growth of the others. The second effect is more subtle and is explained for the two voltage levels 2100 V and 2200 V. The first effect would imply that a voltage level of 2200 V is more suitable on system level; however, this is clearly not the case. While the difference in motor and cable mass for the two voltages is negligible, it is not in terms of inverter efficiency and mass (see Figure 5) as $V_{DC} = 2100$ V represents one of the above-mentioned section boundaries where the selected semiconductor blocking voltage and thus the module changes. This results in a quantifiable effect on system level.

To clarify this, Figure 7b shows the mass breakdown of the single components for the Pareto optimal points at a similar system efficiency $\eta_{Sys} = 94.0\%$ for the four different voltage levels. Motor and inverter masses only feature a minor rise in absolute terms, while cable masses reduce significantly but still have a significant to dominating share (roughly 30–65% of the total mass depending on the voltage level).

It can be concluded that a high voltage level is beneficial in terms of system mass, however not for the electric machine and some inverter topologies, without considering further implications such as available semiconductor modules for even higher voltages or increasing complexity. Possible solutions might be the insertion of DC/DC converters after rectifier and before inverter to reduce the voltage level of motor and inverter and increase the one of the DC transmission system or the use of superconducting technologies [11]. This must be evaluated qualitatively.

5. Conclusions

The possibilities of coupled, analytical models for sizing EP systems have been demonstrated for the CENTRELINE project. A sensitivity study on the impact of the DC link voltage level revealed opposing effects on the component properties while on the system level the trade was clearly dominated by the DC power transmission cable. The underlying effect of discrete steps for the blocking voltages of semiconductor modules on the system performance has been made visible. Two potential technical solutions to decrease the influence of the DC power transmission system were given.

Further work will focus on the modeling of other EP components, especially heat exchangers to account for additional masses due to the thermal management needed for an electric aircraft, as well as refining existing models, e.g., passive masses of electric machines and EMI filters of inverters. Optimization strategies will be implemented.

A complete integration of the presented approach by mirroring back the results of the electric propulsion sizing process into the preliminary aircraft design loop would unlock high potential to improve quality of results as well as to optimize them.

Author Contributions: Conceptualization, S.B., M.B. and M.F.; methodology, S.B., M.B. and M.F.; software, S.B., M.B., S.R. and G.W.; formal analysis, S.B.; investigation, S.B.; data curation, S.B.; writing—original draft preparation, S.B.; writing—review and editing, S.B., M.B. and M.F.; visualization, S.B.; supervision, M.B., M.F. and M.N.; project administration, M.B.; and funding acquisition, G.W. and M.B.

Funding: Part of this work was conducted within the CENTRELINE project, which has received funding from the European Union's Horizon 2020 research and innovation programme under Grant Agreement No. 723242.

Acknowledgments: They authors want to thank their colleagues Andreas Mayr, Christoph Marxguth, Mabroor Ahmed and Torsten Schröder for fruitful discussions.

Abbreviations

The following abbreviations are used in this manuscript:

EP	Electric Propulsion
DOF	Degree of Freedom
DSE	Design Space Exploration
KPI	Key Performance Indicator
DC	Direct Current
AC	Alternating Current
CO_2	Carbon dioxide
NO_x	Nitrogen oxide
NASA	National Aeronautics and Space Exploration
ICAO	International Civil Aviation Organization
ATAG	Aviation Transport Action Group
MDAO	Multidisciplinary Analysis and Optimization
EU	European Union
BLI	Boundary Layer Ingestion
GTF	Geared Turbo Fan
LPT	Low Pressure Turbine
EM	Electric Machine
FF	Fuselage Fan
SiC	Silicon Carbide
PMSM	Permanent Magnet Synchronous Machine
THD	Total Harmonic Distortion
NPC	Neutral Point Clamped

FC   Fly Cap
SMC   Stacked Multi Cell
EMI   Electromagnetic Interference

References

1. Penner, J.E.; Lister, D.H.; Griggs, D.J.; Dokken, D.J.; McFarland, M. *Aviation and the Global Atmosphere: A Special Report of the Intergovernmental Panel on Climate Change*; Cambridge University Press: Cambridge, UK, 1999.
2. Air Transport Action Group. *Aviation: Benefit Beyond Borders*; ATAG: Geneva, Switzerland, 2018.
3. Follen, G.J.; Del Rosario, R.; Wahls, R.; Madavan, N. *NASA's Fundamental Aeronautics Subsonic Fixed Wing Project: Generation N+3 Technology Portfolio*; SAE Technical Paper Series; SAE International: Warrendale, PA, USA, 2011.
4. Guynn, M.D.; Berton, J.J.; Tong, M.J.; Haller, W.J. Advanced Single-Aisle Transport Propulsion Design Options Revisited. In Proceedings of the Aviation Technology, Integration, and Operations Conference, Los Angeles, CA, USA, 12–14 August 2013; American Institute of Aeronautics and Astronautics: Reston, VA, USA, 2013.
5. International Civil Aviation Organization. *Doc 10127: Independent Expert Integrated Technology Goals Assessment and Review for Engines and Aircraft*; ICAO: Montréal, QC, Canada, 2019.
6. International Civil Aviation Organization. *Resolution A39-3: Consolidated Statement of Continuing ICAO Policies and Practices Related to Environmental Protection—Global Market-based Measure (MBM) Scheme*; ICAO: Montréal, QC, Canada, 2016.
7. Darecki, M.; Edelstenne, C.; Enders, T.; Fernandez, E.; Hartman, P.; Herteman, J.P.; Kerkloh, M.; King, I.; Ky, P.; Mathieu, M.; et al. *Flightpath 2050: Europe's Vision for Aviation: Report of the High Level Group*; Publications Office of the European Union: Luxembourg, 2011.
8. Aigner, B.; Nollmann, M.; Stumpf, E. Design of a Hybrid Electric Propulsion System within a Preliminary Aircraft Design Software Environment. In *Deutscher Luft- und Raumfahrtkongress*; Deutsche Gesellschaft für Luft- und Raumfahrt-Lilienthal-Oberth e.V.: Bonn, Germany, 2018; pp. 1–14.
9. Brelje, B.J.; Martins, J.R.R.A. Electric, Hybrid, and Turboelectric Fixed-Wing Aircraft: A Review of Concepts, Models, and Design Approaches. *Prog. Aerosp. Sci.* **2019**, *104*, 1–19. [CrossRef]
10. Voskuijl, M.; van Bogaert, J.; Rao, A.G. Analysis and Design of Hybrid Electric Regional Turboprop Aircraft. *CEAS Aeronaut. J.* **2018**, *9*, 15–25. [CrossRef]
11. Gemin, P.; Kupiszewski, T.; Radun, A.; Pan, Y.; Lai, R.; Zhang, D.; Wang, R.; Wu, X.; Jiang, Y.; Galioto, S.; et al. *Architecture, Voltage, and Components for a Turboelectric Distributed Propulsion Electric Grid (AVC-TeDP)*; NASA Glenn Research Center: Cleveland, OH, USA, 2015.
12. Gesell, H.; Wolters, F.; Plohr, M. System of Turbo Electric and Hybrid Electric Propulsion Systems on a Regional Aircraft. In Proceedings of the 31st Congress of the International Council of the Aeronautical Sciences (ICAS), Horizonte, Brazil, 9–14 September 2018; pp. 9–14.
13. Pornet, C. Electric Drives for Propulsion System of Transport Aircraft. In *New Applications of Electric Drives*; Chomat, M., Ed.; InTech: London, UK, 2015.
14. Isikveren, A.T.; Seitz, A.; Vratny, P.C.; Pornet, C.; Plötner, K.O.; Hornung, M. Conceptual Studies of Universally-Electric Systems Architectures Suitable for Transport Aircraft. In *Deutscher Luft-und Raumfahrt Kongress*; Deutsche Gesellschaft für Luft- und Raumfahrt: Bonn, Germany, 2012.
15. Andrea, J.; Buffo, M.; Guillard, E.; Landfried, R.; Boukadoum, R.; Teste, P. Arcing Fault in Aircraft Distribution Network. In Proceedings of the 2017 IEEE Holm Conference on Electrical Contacts, Denver, CO, USA, 10–13 September 2017; IEEE: Piscataway, NJ, USA, 2017; pp. 317–324.
16. Nya, B.H.; Brombach, J.; Schulz, D. Benefits of Higher Voltage Levels in Aircraft Electrical Power Systems. In Proceedings of the 2012 Electrical Systems for Aircraft, Railway and Ship Propulsion Conference, Bologna, Italy, 16–18 October 2012; pp. 1–5.
17. Jansen, R.; Bowman, C.; Jankovsky, A. Sizing Power Components of an Electrically Driven Tail Cone Thruster and a Range Extender. In Proceedings of the 16th AIAA Aviation Technology, Integration, and Operations Conference, Washington, DC, USA, 13–17 June 2016; p. 3766.
18. Seitz, A. H2020 CENTRELINE—Project Preview. In Proceedings of the 7th EASN International Conference, Warsaw, Poland, 26–29 September 2017.

19. Pornet, C.; Gologan, C.; Vratny, P.C.; Seitz, A.; Schmitz, O.; Isikveren, A.T.; Hornung, M. Methodology for Sizing and Performance Assessment of Hybrid Energy Aircraft. *J. Aircr.* **2015**, *52*, 341–352. [CrossRef]

20. Riboldi, C.E.D.; Gualdoni, F.; Trainelli, L. Preliminary Weight Sizing of Light Pure-Electric and Hybrid-Electric Aircraft. *Transp. Res. Procedia* **2018**, *29*, 376–389. [CrossRef]

21. Stückl, S. Methods for the Design and Evaluation of Future Aircraft Concepts Utilizing Electric Propulsion Systems. Ph.D. Thesis, TU München, Munich, Germany, 2016.

22. Vratny, P.C. Conceptual Design Methods of Electric Power Architectures for Hybrid Energy Aircraft. Ph.D. Thesis, TU München, Munich, Germany, 2019.

23. Vratny, P.C.; Hornung, M. Sizing Considerations of an Electric Ducted Fan for Hybrid Energy Aircraft. *Transp. Res. Procedia* **2018**, *29*, 410–426. [CrossRef]

24. Isikveren, A.T.; Seitz, A.; Bijewitz, J.; Hornung, M.; Mirzoyan, A.; Isyanov, A.; Godard, J.L.; Stückl, S.; van Toor, J. Recent Advances in Airframe-Propulsion Concepts with Distributed Propulsion. In Proceedings of the 29th Congress of the International Council of the Aeronautical Sciences (ICAS), St. Petersburg, Russia, 7–12 September 2014.

25. Meller, F.; Kocvara, F. *Specification of Propulsive Fuselage Aircraft Layout and Design Features: Grant Agreement No. 723242, CENTRELINE Project Deliverable D1.02*; Publications Office of the European Union: Luxembourg, 2018.

26. Seitz, A.; Peter, F.; Bijewitz, J.; Habermann, A.; Goraj, Z.; Kowalski, M.; Castillo, A.; Meller, F.; Merkler, R.; Samuelsson, S.; et al. Concept Validation Study for Fuselage Wake-Filling Propulsion Integration. In Proceedings of the 31st Congress of the International Council of the Aeronautical Sciences (ICAS), Belo Horizonte, Brazil, 9–14 September 2018.

27. Wortmann, G. *Electric Machinery Preliminary Design Report: Grant Agreement No. 723242, CENTRELINE Project Deliverable D4.04*; Publications Office of the European Union: Luxembourg, 2018.

28. Fang, L.C.; Qin, S.Y. Concurrent Optimization for Parameters of Powertrain and Control System of Hybrid Electric Vehicle Based on Multi-Objective Genetic Algorithms. In Proceedings of the 2006 SICE-ICASE International Joint Conference, Busan, Korea, 18–21 October 2006; pp. 2424–2429.

29. Cao, W.; Mecrow, B.C.; Atkinson, G.J.; Bennett, J.W.; Atkinson, D.J. Overview of Electric Motor Technologies Used for More Electric Aircraft (MEA). *IEEE Trans. Ind. Electron.* **2012**, *59*, 3523–3531.

30. Hsieh, M.; Hsu, Y. A Generalized Magnetic Circuit Modeling Approach for Design of Surface Permanent-Magnet Machines. *IEEE Trans. Ind. Electron.* **2012**, *59*, 779–792. [CrossRef]

31. Siemens PLM Software Inc. *Simcenter SPEED: PC-BDC 13.06 User's Manual*; Siemens PLM Software Inc.: Plano, TX, USA, 2018.

32. Boresi, A.P.; Schmidt, R.J. *Advanced Mechanics of Materials*, 6th ed.; Wiley: New York, NY, USA, 2003.

33. Eslami, M.R. Theory of Elasticity and Thermal Stresses: Explanations, Problems and Solutions. In *Solid Mechanics and Its Applications*; Springer: Dordrecht, The Netherlands, 2013; Volume 197.

34. IEC 60664-1:2007. *Insulation Coordination for Equipment within Low-Voltage Systems*; VDE-Verlag: Berlin, Germany, 2007.

35. Golovanov, D.; Papini, L.; Gerada, D.; Xu, Z.; Gerada, C. Multidomain optimization of High-Power-Density PM Electrical Machines for System Architecture Selection. *IEEE Trans. Ind. Electron.* **2018**, *65*, 5302–5312. [CrossRef]

36. Krug, D. Vergleichende Untersuchungen von Mehrpunkt-Schaltungstopologien mit zentralem Gleich-Spannungszwischenkreis für Mittelspannungsanwendungen. (German) [Comparison of Medium-Voltage Multilevel Converters with Central DC Link, Chap. 3]. Ph.D. Thesis, TU Dresden, Dresden, Germany, 2016.

37. Krug, D.; Bernet, S.; Fazel, S.S.; Jalili, K.; Malinowski, M. Comparison of 2.3-kV Medium-Voltage Multilevel Converters for Industrial Medium-Voltage Drives. *IEEE Trans. Ind. Electron.* **2007**, *54*, 2979–2992. [CrossRef]

38. Brückner, T. The Active NPC Converter for Medium Voltage Drives. Ph.D. Thesis, RWTH Aachen, Aachen, Germany, 2006.

39. Wintrich, A.; Nicolai, U.; Tursky, W.; Reimann, T. *Application Manual Power Semiconductors*; ISLE Verlag: Ilmenau, Germany, 2011.

40. Ilgevicius, A. Analytical and Numerical Analysis and Simulaiton of Heat Transfer in Electrical Conductors and Fuses. Ph.D. Thesis, Universität der Bundeswehr München, Neubiberg, Germany, 2014.

41. Van der Geest, M.; Polinder, H.; Ferreira, J.A.; Christmann, M. Power Density Limits and Design Trends of High-Speed Permanent Magnet Synchronous Machines. *IEEE Trans. Transp. Electrif.* **2015**, *1*, 266–276. [CrossRef]
42. Kolar, J.W.; Drofenik, U.; Biela, J.; Heldwein, M.L.; Ertl, H.; Friedli, T.; Round, S.D. PWM Converter Power Density Barriers. In Proceedings of the Power Conversion Conference, Nagoya, Japan, 2–5 April 2007; pp. 9–29.
43. Coroplast GmbH & Co. KG. *Wires and Cables for Automotive Applications*; Coroplast GmbH & Co. KG: Wuppertal, Germany, 2016.

MDPI
St. Alban-Anlage 66
4052 Basel
Switzerland
Tel. +41 61 683 77 34
Fax +41 61 302 89 18
www.mdpi.com

Aerospace Editorial Office
E-mail: aerospace@mdpi.com
www.mdpi.com/journal/aerospace